GCSE
IN A
WEEK

Maths to A*

Fiona Mapp

Revision Planner

DAY 5

Page	Time	Topic	Date	Time	Completed
52	15 minutes	Loci			
54	20 minutes	Translations and Reflections			
56	15 minutes	Rotation and Enlargement			
58	15 minutes	Similarity and Congruency 1			
60	20 minutes	Similarity and Congruency 2			
62	20 minutes	Circle Theorems			
64	15 minutes	Pythagoras' Theorem			

DAY 6

Page	Time	Topic	Date	Time	Completed
66	20 minutes	Trigonometry			
68	15 minutes	Trigonometry in 3D			
70	25 minutes	Sine and Cosine Rules			
72	15 minutes	Trigonometric Functions			
74	15 minutes	Arc, Sector and Segment			
76	20 minutes	Surface Area and Volume ✓			
78	15 minutes	Vectors 1			

DAY 7

Page	Time	Topic	Date	Time	Completed
80	20 minutes	Vectors 2			
82	15 minutes	Scatter Diagrams and Correlation			
84	15 minutes	Averages			
86	25 minutes	Cumulative Frequency Graphs			
88	20 minutes	Histograms			
90	25 minutes	Probability			

92 Answers

810 Mins
13:30 hours
(27 half an hours)

Prime Factors, HCF and LCM

Prime Factors

Apart from **prime numbers**, any whole number greater than 1 can be written as a product of **prime factors**. This means the number is written using only prime numbers multiplied together.

A prime number has only two factors, 1 and itself. 1 is not a prime number.

The prime numbers up to 20 are:

2, 3, 5, 7, 11, 13, 17, 19

The diagram below shows the prime factors of 60.

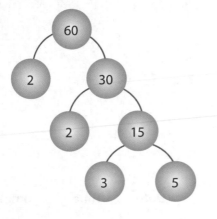

⚫ Divide 60 by its first prime factor, 2.

⚫ Divide 30 by its first prime factor, 2.

⚫ Divide 15 by its first prime factor, 3.

⚫ We can now stop because the number 5 is prime.

As a product of its prime factors, 60 may be written as:

$60 = 2 \times 2 \times 3 \times 5$

or

$60 = 2^2 \times 3 \times 5$

Highest Common Factor (HCF)

The highest factor that two numbers have in common is called the **HCF**.

Example

Find the HCF of 60 and 96.

⚫ Write the numbers as products of their prime factors.

$60 = 2 \times 2 \qquad \times 3 \times 5$

$96 = 2 \times 2 \times 2 \times 2 \times 2 \times 3$

⚫ Ring the factors that are common.

$60 = 2 \times 2 \qquad \times 3 \times 5$

$96 = 2 \times 2 \times 2 \times 2 \times 2 \times 3$

⚫ These give the HCF $= 2 \times 2 \times 3$

$\qquad\qquad\qquad = \mathbf{12}$

Lowest (Least) Common Multiple (LCM)

The **LCM** is the lowest number that is a multiple of two numbers.

Example
Find the LCM of 60 and 96.

- Write the numbers as products of their prime factors.

    $60 = 2 \times 2 \qquad \times 3 \times 5$

    $96 = 2 \times 2 \times 2 \times 2 \times 2 \times 3$

- 60 and 96 have a common factor of $2 \times 2 \times 3$, so it is only counted once.

    $60 = \boxed{2} \times \boxed{2} \qquad \times \boxed{3} \times 5$

    $96 = \boxed{2} \times \boxed{2} \times 2 \times 2 \times 2 \times \boxed{3}$

- The LCM of 60 and 96 is

    $2 \times 2 \times 2 \times 2 \times 2 \times 3 \times 5$

    $= \textbf{480}$

- Any whole number greater that 1 can be written as a product of its prime factors, apart from prime numbers themselves (1 is not prime).

- The highest factor that two numbers have in common is called the Highest Common Factor (HCF).

- The lowest number that is a multiple of two numbers is called the Lowest (Least) Common Multiple (LCM).

QUESTIONS

QUICK TEST

1. Write these numbers as products of their prime factors:

    a. 50

    b. 360

    c. 16

2. Decide whether these statements are true or false.

    a. The HCF of 20 and 40 is 4.

    b. The LCM of 6 and 8 is 24.

    c. The HCF of 84 and 360 is 12.

    d. The LCM of 24 and 60 is 180.

EXAM PRACTICE

1. Find the Highest Common Factor of 120 and 42.

2. Buses to St Albans leave the bus station every 20 minutes. Buses to Hatfield leave the bus station every 14 minutes.

    A bus to St Albans and a bus to Hatfield both leave the bus station at 10 am. When will buses to both St Albans and Hatfield next leave the bus station at the same time?

Fractions and Recurring Decimals

A **fraction** is part of a whole number. The top number is the **numerator** and the bottom number is the **denominator**.

There are four rules of fractions.

Addition $\boxed{+}$

You need to change the fractions so that they have the same denominator.

Example

$$\frac{5}{9} + \frac{1}{7}$$

The lowest common denominator is 63, since both 9 and 7 go into 63.

$$= \frac{35}{63} + \frac{9}{63}$$

Remember to only add the numerators and not the denominators.

$$= \frac{44}{63}$$

Subtraction $\boxed{-}$

You need to change the fractions so that they have the same denominator.

Example

$$\frac{4}{5} - \frac{1}{3}$$

The lowest common denominator is 15.

$$= \frac{12}{15} - \frac{5}{15}$$

Remember to only subtract the numerators and not the denominators.

$$= \frac{7}{15}$$

Multiplication $\boxed{\times}$

Before starting, write out whole or mixed numbers as improper fractions (also known as top-heavy fractions).

Example

$$\frac{2}{7} \times \frac{4}{5}$$

Multiply the numerators together.

$$= \frac{2 \times 4}{7 \times 5}$$

Multiply the denominators together.

$$= \frac{8}{35}$$

Division $\boxed{\div}$

Before starting, write out whole or mixed numbers as improper fractions (also known as top-heavy fractions).

Example

$$2\frac{1}{3} \div 1\frac{2}{7}$$

$$= \frac{7}{3} \div \frac{9}{7}$$ ← Convert to top-heavy fractions.

$$= \frac{7}{3} \times \frac{7}{9}$$ ← Take the reciprocal of the second fraction and multiply both fractions.

$$= \frac{49}{27}$$

$$= \mathbf{1\frac{22}{27}}$$ ← Rewrite the fraction as a mixed number.

Fraction Problems

You may need to solve problems involving fractions.

Example
Charlotte's take-home pay is £930. She gives her mother $\frac{1}{3}$ of this and spends $\frac{1}{5}$ of the £930 on going out. What fraction of the £930 is left?

Give your answer as a fraction in its simplest form.

$$\frac{1}{3} + \frac{1}{5}$$

This is a simple addition of fractions question.

$$= \frac{5}{15} + \frac{3}{15}$$

Write the fractions with a common denominator.

$$= \frac{8}{15}$$

$$1 - \frac{8}{15}$$

The question asks for the fraction of the money that is left, so subtract $\frac{8}{15}$ from 1.

$$= \frac{7}{15}$$

The fraction is in its simplest form.

Changing Recurring Decimals to Fractions

Recurring decimals are **rational** numbers because we can change them to fractions.

Examples

1. Change $0.\dot{1}\dot{5}$ to a fraction in its lowest terms.

   Let $x = 0.151515 \ldots$ ①

   then $100x = 15.151515 \ldots$ ②

   > Multiply by 10^n, where n is the length of the recurring pattern.

   Subtract equation ① from equation ②.

   ② – ① $99x = 15$

   $$x = \frac{15}{99}$$

   $$x = \frac{5}{33}$$ | Check $5 \div 33 = 0.\dot{1}\dot{5}$

2. Change $0.3\dot{7}$ into a fraction.

   $$x = 0.3777 \ldots \text{①}$$

   $$10x = 3.7777 \ldots \text{②}$$

   > Multiply by 10, so that the part that is not recurring is in the units position.

   $$100x = 37.7777 \ldots \text{③}$$

   > Multiply equation ① by 100.

   Subtract equation ② from equation ③.

   ③ – ② $90x = 34$

   $$x = \frac{34}{90}$$

   $$x = \frac{17}{45}$$ | Always check to see if your fraction simplifies.

SUMMARY

- To add or subtract fractions, write them using the same denominator.
- To multiply fractions, multiply the numerators and multiply the denominators.
- To divide fractions, take the reciprocal of the second fraction and multiply the fractions together.
- When multiplying and dividing fractions, write out whole or mixed numbers as top-heavy fractions before you begin the calculation.
- Recurring decimals are rational numbers, so they can be changed to fractions.

QUESTIONS

QUICK TEST

1. Work out the following:

   a. $\frac{2}{3} + \frac{1}{5}$

   b. $2\frac{6}{7} - \frac{1}{3}$

   c. $\frac{2}{9} \times \frac{5}{7}$

   d. $\frac{3}{11} \div \frac{22}{27}$

2. Change the following recurring decimals into fractions. Write each fraction in its simplest form.

   a. $0.\dot{7}$

   b. $0.\dot{2}1\dot{5}$

   c. $0.3\dot{5}$

EXAM PRACTICE

1. In a magazine $\frac{3}{7}$ of the pages have advertisements on them. Given that 12 pages have advertisements on them, work out the number of pages in the magazine.

2. Prove that $0.\dot{3}\dot{6}$ is equivalent to $\frac{4}{11}$.

Repeated Percentage Change

Percentage Change

> Percentage change = $\frac{\text{change}}{\text{original}} \times 100\%$

Examples

1. Tammy bought a flat for £185 000. Three years later she sold it for £242 000. What is her percentage profit?

   Profit is £242 000 − £185 000

   $\qquad$ = £57 000

   Percentage profit is $\frac{57\,000}{185\,000} \times 100\%$

   $\qquad\qquad$ = **30.8%** (3sf)

2. Jackie bought a car for £12 500 and sold it two years later for £7250. Work out her percentage loss.

   Loss is £12 500 − £7250

   $\qquad$ = £5250

   Percentage loss is $\frac{5250}{12\,500} \times 100\%$

   $\qquad\qquad$ = **42%**

Repeated Percentage Change

A quantity can increase or decrease in value each year by a different percentage. These quantities will change in value at the end of each year. To calculate repeated percentage change, two methods are explained in the example below.

Example

A car was bought for £12 500. Each year it depreciated in value by 15%. What was the car worth after three years?

> You must remember **not** to do 3 × 15% = 45% reduction over 3 years!

Method 1

● Find 100% − 15% = 85% of the value of the car first.

   Year 1: $\frac{85}{100} \times$ £12 500 = £10 625

● Then work out the value year by year. (£10 625 depreciates in value by 15%.)

   Year 2: $\frac{85}{100} \times$ £10 625 = £9031.25

   (£9301.25 depreciates in value by 15%.)

   Year 3 : $\frac{85}{100} \times$ £9031.25 = **£7676.56** (2dp)

Method 2

● A quick way to work this out is by using a multiplier.

● Finding 85% of the value of the car is the same as multiplying by 0.85

   Year 1: 0.85 × £12 500 = £10 625

   Year 2: 0.85 × £10 625 = £9031.25

   Year 3: 0.85 × £9031.25 = **£7676.56** (2dp)

● This is the same as working out $(0.85)^3 \times$ £12 500 = **£7676.56** (2dp)

Compound Interest

Compound interest is where the bank pays interest on the interest already earned as well as on the original money.

Example
Becky has £3200 in her savings account and compound interest is paid at 3.2% per annum. How much will she have in her account after four years?

100% + 3.2% = 103.2%

= 1.032 — This is the multiplier.

Year 1: 1.032 × £3200 = £3302.40

Year 2: 1.032 × £3302.40 = £3408.08

Year 3: 1.032 × £3408.08 = £3517.14

Year 4: 1.032 × £3517.14 = £3629.68

Total = **£3629.68** (2dp)

A quicker way is to multiply £3200 by $(1.032)^4$

Number of years

£3200 × $(1.032)^4$ = **£3629.68** (2dp)

Original

Multiplier

Simple Interest

Simple interest is the interest paid each year. It is the same amount each year.

The simple interest on £3200 invested for four years at 3.2% per annum would be:

$\frac{3.2}{100}$ × 3200 = £102.40 for one year

Interest over four years would be 4 × £102.40 = £409.60

Total in account after 4 years would be £3609.60

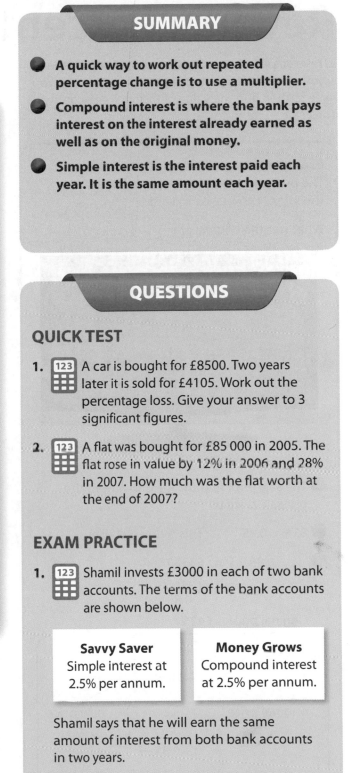

- A quick way to work out repeated percentage change is to use a multiplier.
- Compound interest is where the bank pays interest on the interest already earned as well as on the original money.
- Simple interest is the interest paid each year. It is the same amount each year.

QUESTIONS

QUICK TEST

1. `123` A car is bought for £8500. Two years later it is sold for £4105. Work out the percentage loss. Give your answer to 3 significant figures.

2. `123` A flat was bought for £85 000 in 2005. The flat rose in value by 12% in 2006 and 28% in 2007. How much was the flat worth at the end of 2007?

EXAM PRACTICE

1. `123` Shamil invests £3000 in each of two bank accounts. The terms of the bank accounts are shown below.

Savvy Saver	**Money Grows**
Simple interest at 2.5% per annum.	Compound interest at 2.5% per annum.

Shamil says that he will earn the same amount of interest from both bank accounts in two years.

Decide whether Shamil is correct. You must show full working to justify your answer.

Reverse Percentage Problems

In reverse percentage problems you are given the final amount after a percentage increase or decrease. You have to then find the value of the original quantity. These are quite tricky, so think carefully.

Example 1

The price of a television is reduced by 15% in the sales. It now costs £352.75

What was the original price?

- The sale price is 100% – 15% = 85% of the pre-sale price (x)

- 85% = 0.85   This is the multiplier.

- $0.85 \times x = £352.75$

$$x = \frac{£352.75}{0.85}$$

Original price is **£415**

Check:

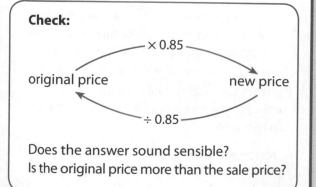

original price — × 0.85 → new price

new price — ÷ 0.85 → original price

Does the answer sound sensible?
Is the original price more than the sale price?

Example 2

A telephone bill costs £169.20 including VAT at 20%. What is the cost of the bill without the VAT?

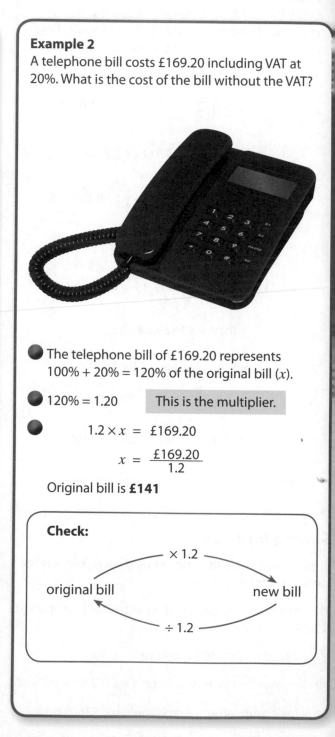

- The telephone bill of £169.20 represents 100% + 20% = 120% of the original bill (x).

- 120% = 1.20   This is the multiplier.

- $1.2 \times x = £169.20$

$$x = \frac{£169.20}{1.2}$$

Original bill is **£141**

Check:

original bill — × 1.2 → new bill

new bill — ÷ 1.2 → original bill

Example 3

The price of a washing machine is reduced by 5% in the sales. It now costs £323. What was the original price?

5% off

- The sale price is 100% – 5% = 95% of the pre-sale price (x).

- 95% = 0.95   This is the multiplier.

- $0.95 \times x = £323$

  $x = \dfrac{£323}{0.95}$

Original price is **£340**

Check:

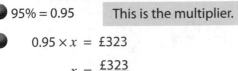

× 0.95

original price ⟶ new price

÷ 0.95

SUMMARY

- A reverse percentage problem is where you know the amount after a percentage change and want to find the original amount.

- Use a multiplier to work out reverse percentage problems.

- Always check that your answer seems sensible.

QUESTIONS

QUICK TEST

1. [123] Each item listed below includes VAT at 20%. Work out the original price of the item.

   a. A pair of shoes: £69

   b. A coat: £152.40

   c. A suit: £285

   d. A television: £525

EXAM PRACTICE

1. [123] In a sale, normal prices are reduced by 12%. The sale price of a television is £220.

   Work out the normal price of the television.

2. [123] Joseph says that the original price of a CD player, which now costs £60 after a 15% reduction, was £70.59

   Is Joseph correct? Show your working.

Ratio and Proportion

Sharing a Quantity in a Given Ratio

A **ratio** is used to compare two or more related quantities.

To share an amount in a given ratio, add up the individual parts and then divide the amount by this number to find one part.

Example

£155 is divided in the ratio of 2 : 3 between Daisy and Tom. How much does each receive?

$2 + 3 = 5$ parts — Add up the total parts.

5 parts = £155

1 part = £155 ÷ 5 — Work out what one part is worth.

= £31

So Daisy gets $2 \times £31 =$ **£62**
and Tom gets $3 \times £31 =$ **£93**

Check: £62 + £93
= £155 ✔

Exchange Rates

Two quantities are in **direct proportion** when both quantities increase at the same rate.

Example

Samuel went on holiday to Spain. He changed £350 into Euros. The exchange rate was £1 = €1.16. How many Euros did Samuel receive?

£1 = €1.16 so

£350 = 350 × 1.16

= **€406**

Best Buys

Use unit amounts to help you decide which is the better value for money.

Example

The same brand of breakfast cereal is sold in two different-sized packets. Which packet represents better value for money?

£3.15

£1.65

500g

125g

Find the cost per gram for both boxes of cereal.

125 g costs £1.65 so $\frac{165}{125} = 1.32$p per gram

500 g costs £3.15 so $\frac{315}{500} = 0.63$p per gram

Since the larger box costs less per gram, it represents better value for money.

ncreasing and Decreasing in a Given Ratio

When increasing or decreasing in a given ratio, it is sometimes easier to find a unit amount.

Example

A photograph of length 12 cm is to be enlarged in the ratio 4 : 5

What is the length of the enlarged photograph?

$\frac{12}{4} = 3$ cm

Divide 12 by 4 to get 1 part.

$3 \times 5 = \mathbf{15}$ cm

Multiply this by 5 to get the length of the enlarged photograph.

When two quantities are in **inverse proportion**, one quantity increases at the same rate as the other quantity decreases. For example, the time it takes to build a wall increases as the number of builders decreases.

A wall took 4 builders 6 days to build.

Time for 4 builders is 6 days

Time for 1 builder is $6 \times 4 = 24$ days

It takes 1 builder four times as long to build the wall.

At the same rate it would take 6 builders $\frac{24}{6} = 4$ days

SUMMARY

- A ratio is used to compare two or more related quantities.

- Two quantities are in direct proportion when both quantities increase at the same rate.

- Two quantities are in inverse proportion when one quantity increases at the same rate as the other quantity decreases.

QUESTIONS

QUICK TEST

1. Divide £160 in the ratio 1 : 2 : 5

2. 123 The cost of four ring binders is £6.72
   Work out the cost of 21 ring binders.

3. A patio took 6 builders 4 days to lay. At the same rate how long would it take 8 builders?

EXAM PRACTICE

1. Toothpaste is sold in three different-sized tubes.

   50 ml is £1.24
   75 ml is £1.96
   100 ml is £2.42

   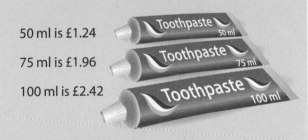

   Which of the tubes of toothpaste is the better value for money? You must show full working in order to justify your answer.

2. Jessica buys a pair of jeans in England for £52. She then goes on holiday to America and sees an identical pair of jeans for $63. The exchange rate is £1 = $1.49. In which country are the jeans cheaper, and by how much?

Indices

An **index** is sometimes called a **power**.

The base → a^b ← The index or power

Laws of Indices

The laws of indices can be used for numbers or in algebra. The base has to be the same when the laws of indices are applied.

$$a^n \times a^m = a^{n+m}$$

$$a^n \div a^m = a^{n-m}$$

$$(a^n)^m = a^{n \times m}$$

$$a^0 = 1$$

$$a^1 = a$$

$$a^{-n} = \frac{1}{a^n}$$

$$a^{\frac{1}{m}} = \sqrt[m]{a}$$

$$a^{\frac{n}{m}} = (\sqrt[m]{a})^n$$

Examples with Numbers

1. Simplify the following, leaving your answers in index notation.

   a. $5^2 \times 5^3 = 5^{2+3} = \mathbf{5^5}$

   b. $8^{-5} \times 8^{12} = 8^{-5+12} = \mathbf{8^7}$

   c. $(2^3)^4 = 2^{3\times4} = \mathbf{2^{12}}$

2. Evaluate: *Evaluate means to work out.*

   a. $4^2 = 4 \times 4 = \mathbf{16}$

   b. $5^0 = \mathbf{1}$

   c. $3^{-2} = \frac{1}{3^2} = \mathbf{\frac{1}{9}}$

   d. $36^{\frac{1}{2}} = \sqrt{36} = \mathbf{6}$

   e. $8^{\frac{2}{3}} = (\sqrt[3]{8})^2 = 2^2 = \mathbf{4}$

3. Simplify the following, leaving your answers in index form.

   a. $7^2 \times 7^5 = \mathbf{7^7}$

   b. $6^9 \div 6^2 = \mathbf{6^7}$

   c. $\frac{3^7 \times 3^2}{3^{10}} = \frac{3^9}{3^{10}} = \mathbf{3^{-1}}$

   d. $7^9 \div 7^{-10} = \mathbf{7^{19}}$

4. Evaluate:

   a. $3^3 = 3 \times 3 \times 3 = \mathbf{27}$

   b. $7^0 = \mathbf{1}$

   c. $64^{\frac{1}{3}} = \sqrt[3]{64} = \mathbf{4}$

   d. $81^{\frac{1}{2}} = \sqrt{81} = \mathbf{9}$

   e. $5^{-2} = \frac{1}{5^2} = \mathbf{\frac{1}{25}}$

   f. $\left(\frac{4}{9}\right)^{-2} = \left(\frac{9}{4}\right)^2 = \frac{81}{16} = \mathbf{5\frac{1}{16}}$

Examples with Algebra

1. Simplify the following:

a. $a^4 \times a^{-6} = a^{4-6} = a^{-2} = \dfrac{1}{a^2}$

b. $5y^2 \times 3y^6 = \mathbf{15y^8}$

| The numbers are multiplied. | The indices are added. |

c. $(4x^3)^2 = \mathbf{16x^6}$

Remember to square the 4 as well.

If in doubt, write it out: $(4x^3)^2 = 4x^3 \times 4x^3$

$= \mathbf{16x^6}$

d. $(3x^4y^2)^3 = \mathbf{27x^{12}y^6}$

or $3x^4y^2 \times 3x^4y^2 \times 3x^4y^2 = \mathbf{27x^{12}y^6}$

e. $(2x)^{-3} = \dfrac{1}{(2x)^3} = \dfrac{1}{\mathbf{8x^3}}$

2. Simplify:

a. $\dfrac{15b^4 \times 3b^7}{5b^2} = \dfrac{45b^{11}}{5b^2} = \mathbf{9b^9}$

b. $\dfrac{16a^2b^4}{4ab^3} = \mathbf{4ab}$

3. Simplify:

a. $7a^2 \times 3a^2b = \mathbf{21a^4b}$

b. $\dfrac{14a^2b^4}{7ab} = \mathbf{2ab^3}$

c. $\dfrac{9x^2y \times 2xy^3}{6xy} = \dfrac{18x^3y^4}{6xy}$

$= \mathbf{3x^2y^3}$

SUMMARY

● Make sure you know and can use all the laws of indices.

● A negative power is the reciprocal of the positive power.

● Fractional indices mean roots.

QUESTIONS

QUICK TEST

1. Simplify the following, leaving your answers in index form.

a. $6^3 \times 6^5$

b. $12^{10} \div 12^{-3}$

c. $(5^2)^3$

d. $64^{\frac{2}{3}}$

2. Simplify the following:

a. $2b^4 \times 3b^6$

b. $8b^{-12} \div 4b^4$

c. $(3b^4)^2$

d. $(5x^2y^3)^{-2}$

EXAM PRACTICE

1. Evaluate:

a. 5^0 b. 7^{-2}

c. $64^{\frac{1}{3}}$ d. $27^{-\frac{2}{3}}$

2. Simplify:

a. $\dfrac{x^4 \times x^7}{x^{15}}$

b. $\dfrac{3x^4 \times 4x^2}{2x^3}$

Standard Index Form

Standard index form (standard form) is useful for writing very large or very small numbers in a simpler way.

When written in standard form a number will be written as:

A number between 1 and 10 $1 \leqslant a < 10$	→ $a \times 10^n$

The value of n is the number of places the digits have to be moved to return the number to its original value.

If the number is 10 or more, n is positive.

If the number is less than 1, n is negative.

If the number is between 1 and 10, n is zero.

Examples

1. Write 2 730 000 in standard form.

   ● 2.73 is the number between 1 and 10 ($1 \leqslant 2.73 < 10$)

   ● Count how many spaces the digits have to move to restore the original number.

   The digits have moved 6 places to the left because it has been multiplied by 10^6

   2 . 7 3

   2 7 3 0 0 0 0

   So, 2 730 000 = **2.73 × 10⁶**

2. Write 0.000046 in standard form.

   ● Put the decimal point between the 4 and 6, so the number lies between 1 and 10.

   ● Move the digits five places to the right to restore the original number.

   ● The value of n is negative.

   So, 0.000046 = **4.6 × 10⁻⁵**

On a Calculator

 To put a number written in standard form into your calculator you use the following key:

$$\boxed{\text{EXP}} \quad \boxed{\text{EE}} \quad \text{or} \quad \boxed{\times 10^x}$$

For example, $(2 \times 10^3) \times (6 \times 10^7) = 1.2 \times 10^{11}$ would be keyed in as:

$$\boxed{2}\ \boxed{\text{EXP}}\ \boxed{3}\ \boxed{\times}\ \boxed{6}\ \boxed{\text{EXP}}\ \boxed{7}\ \boxed{=}$$

or $\boxed{2}\ \boxed{\times 10^x}\ \boxed{3}\ \boxed{\times}\ \boxed{6}\ \boxed{\times 10^x}\ \boxed{7}\ \boxed{=}$

Doing Calculations

Examples

 Work out the following using a calculator. Check that you get the answers given here.

1. $(6.7 \times 10^7)^3 = \mathbf{3.0 \times 10^{23}}$ (2sf)

2. $\dfrac{(4 \times 10^9)}{(3 \times 10^4)^2} = \mathbf{4.\dot{4}}$

3. $\dfrac{(5.2 \times 10^6) \times (3 \times 10^7)}{(4.2 \times 10^5)^2} = \mathbf{884.4}$ (1dp)

Examples

On a non-calculator paper you can use indices to help work out your answers.

1. $(2 \times 10^3) \times (6 \times 10^7)$

$= (2 \times 6) \times (10^3 \times 10^7)$

$= 12 \times 10^{3+7}$

$= 12 \times 10^{10}$

$= 1.2 \times 10^1 \times 10^{10}$

$= \mathbf{1.2 \times 10^{11}}$

2. $(6 \times 10^4) \div (3 \times 10^{-2})$

$= (6 \div 3) \times (10^4 \div 10^{-2})$

$= 2 \times 10^{4-(-2)}$

$= \mathbf{2 \times 10^6}$

3. $(3 \times 10^4)^2$

$= (3 \times 10^4) \times (3 \times 10^4)$

$= (3 \times 3) \times (10^4 \times 10^4)$

$= \mathbf{9 \times 10^8}$

You also need to be able to work out more complex calculations.

Example

The mass of Saturn is 5.7×10^{26} tonnes. The mass of the Earth is 6.1×10^{21} tonnes. How many times heavier is Saturn than the Earth? Give your answer in standard form, correct to 2 significant figures.

$$\frac{5.7 \times 10^{26}}{6.1 \times 10^{21}} = 93\,442.6$$

Now rewrite your answer in standard form.

Saturn is 9.3×10^4 times heavier than the Earth.

QUESTIONS

QUICK TEST

1. Write in standard form:

   a. 64 000

   b. 0.00046

2. Without a calculator, work out the following. Leave in standard form.

   a. $(3 \times 10^4) \times (4 \times 10^6)$

   b. $(6 \times 10^{-5}) \div (3 \times 10^{-4})$

3. 🖩 Work these out on a calculator:

   a. $(4.6 \times 10^{12}) \div (3.2 \times 10^{-6})$

   b. $(7.4 \times 10^9)^2 + (4.1 \times 10^{11})$

EXAM PRACTICE

1. a. Write 40 000 000 in standard form.

   b. Write 6×10^{-5} as an ordinary number.

2. The mass of an atom is 2×10^{-23} grams. What is the total mass of 7×10^{16} of these atoms?

Give your answer in standard form.

Proportionality

Direct Proportion

In **direct proportion**, as one variable increases the other increases, and as one variable decreases the other decreases.

Example 1

If a is proportional to the square of b and $a = 5$ when $b = 4$, find the value of k (the constant of proportionality) and the value of a when $b = 8$.

⚫ Change the sentence by adding the symbol $\propto$, which means 'is directly proportional to'.

$a \propto b^2$

⚫ Replace $\propto$ with '$= k$' to make an equation.

$a = kb^2$

⚫ Substitute the values given in the question in order to find k.

$5 = k \times 4^2$

Rearrange the equation.

$$\frac{5}{16} = k$$

⚫ Replace k with the value just found.

$a = \frac{5}{16}b^2$

If $b = 8$ $a = \frac{5}{16} \times 8^2$

$a = \frac{5}{16} \times 64$

$a = \mathbf{20}$

Example 2

A train is accelerating out of a station at a constant rate.

The distance, d metres, travelled from the station varies directly as the square of the time taken, t seconds.

After 2.5 seconds the train has travelled 20 m.

a. Work out the formula connecting d and t.

$d \propto t^2$

$d = kt^2$

$20 = k \times 2.5^2$

$k = \frac{20}{2.5^2}$

$k = 3.2$

$d = \mathbf{3.2}t^2$

b. How long does the train take to travel 100 m?

$100 = 3.2t^2$

$t^2 = \frac{100}{3.2}$

$t^2 = 31.25$

$t = \sqrt{31.25}$

$t = \mathbf{5.59 \ seconds}$ (3sf)

Inverse Proportion

In **inverse proportion**, as one variable increases the other decreases, and as one variable decreases the other increases.

If y is inversely proportional to x, then we write this as

$$y \propto \frac{1}{x} \text{ or } y = \frac{k}{x}$$

Example

p is inversely proportional to the cube of w.
If $w = 2$ when $p = 5$, what is the value of w when $p = 10$? Give your answer to 3 decimal places.

$p \propto \dfrac{1}{w^3}$ — Write the information with the proportionality sign.

$p = \dfrac{k}{w^3}$ — Replace with the constant of proportionality.

$5 = \dfrac{k}{2^3}$

$5 \times 8 = k$

$k = 40$ — Find the value of k.

$p = \dfrac{40}{w^3}$ — Rewrite the formula.

$10 = \dfrac{40}{w^3}$ — Find the value of w if $p = 10$.

$w^3 = 4$

$w = \sqrt[3]{4}$

$w = \mathbf{1.587}$ (3dp)

SUMMARY

- The notation $\propto$ means 'is directly proportional to'. This is often abbreviated to 'is proportional to' or 'varies as'.

- In direction proportion, if one variable increases the other variable also increases and if one variable decreases, the other variable also decreases.

- In inverse proportion, if one variable increases the other variable decreases.

QUESTIONS

QUICK TEST

1. Match these statements with the correct equation.

a. y is proportional to x	$y = \frac{k}{x}$
b. y is inversely proportional to cube root of x	$y = kx^3$
c. y is inversely proportional to x	$y = kx$
d. y is proportional to x^3	$y = \frac{k}{\sqrt[3]{x}}$

EXAM PRACTICE

1. a is directly proportional to $\sqrt{x}$. When $x = 4$, $a = 8$. What is the equation of proportionality and what is the value of x when $a = 64$?

2. 〔123〕 y is inversely proportional to the square of x. When $x = 3$, $y = 10$.

   a. Calculate y when $x = 2$.

   b. Calculate x when $y = 6$.

Surds

Surds

A **rational number** is one that can be expressed in the form $\frac{a}{b}$, where a and b are integers and $b \neq 0$.

Rational numbers include $\frac{1}{3}$, $0.\dot{7}$, $\sqrt{16}$, $\sqrt[3]{8}$, etc.

Irrational numbers cannot be expressed as a fraction $\frac{a}{b}$.

Irrational numbers include π, π^2, $\sqrt{2}$, $\sqrt{7}$, etc.

> Roots, such as square roots or cube roots that are irrational, are also called **surds**.

Manipulating Surds

When working with surds there are several rules to learn:

1. $\sqrt{a} \times \sqrt{b} = \sqrt{ab}$

Example: $\sqrt{3} \times \sqrt{5} = \sqrt{15}$

2. $(\sqrt{b})^2 = \sqrt{b} \times \sqrt{b} = b$

Example: $(\sqrt{5})^2 = \sqrt{5} \times \sqrt{5} = 5$

3. $\frac{\sqrt{a}}{\sqrt{b}} = \sqrt{\frac{a}{b}}$

Example: $\frac{\sqrt{10}}{\sqrt{2}} = \sqrt{\frac{10}{2}} = \sqrt{5}$

4. $(a + \sqrt{b})^2$

$= (a + \sqrt{b})(a + \sqrt{b})$

$= a^2 + 2a\sqrt{b} + (\sqrt{b})^2$

$= a^2 + 2a\sqrt{b} + b$

5. $(a + \sqrt{b})(a - \sqrt{b})$

$= a^2 - a\sqrt{b} + a\sqrt{b} - (\sqrt{b})^2$

$= a^2 - b$

Examples

1. Simplify $\sqrt{75}$.

$\sqrt{75} = \sqrt{25} \times \sqrt{3}$

$\sqrt{75} = \mathbf{5\sqrt{3}}$

> Look for the highest perfect square, i.e. 25.

2. Expand and simplify $(\sqrt{3} + 2)^2$.

$(\sqrt{3} + 2)(\sqrt{3} + 2)$

$= \sqrt{9} + 2\sqrt{3} + 2\sqrt{3} + 4$

$= \mathbf{7 + 4\sqrt{3}}$

3. Work out $\frac{(2 - \sqrt{2})(4 + 3\sqrt{2})}{2}$. Leave your answer in surd form.

$\frac{(2 - \sqrt{2})(4 + 3\sqrt{2})}{2}$

$= \frac{8 + 6\sqrt{2} - 4\sqrt{2} - 3(\sqrt{2})^2}{2}$

$= \frac{8 + 2\sqrt{2} - 6}{2}$

$= \frac{2 + 2\sqrt{2}}{2}$

$= \frac{2(1 + \sqrt{2})}{2}$

$= \mathbf{1 + \sqrt{2}}$

Rationalising the Denominator

Sometimes a surd can appear on the bottom of the fraction. It is usual to rewrite the surd so that it appears as the numerator. This is called **rationalising** the denominator.

Examples

1. Rationalise $\dfrac{4}{\sqrt{7}}$.

$\dfrac{4}{\sqrt{7}}$

$= \dfrac{4}{\sqrt{7}} \times \dfrac{\sqrt{7}}{\sqrt{7}}$

Multiply the numerator and denominator of the fraction by the surd function. In this case $\sqrt{7}$.

$= \dfrac{4\sqrt{7}}{(\sqrt{7})^2}$

$= \dfrac{4\sqrt{7}}{7}$

2. Given that $\dfrac{4 - \sqrt{18}}{\sqrt{2}} = a + b\sqrt{2}$, where a and b are integers, find the value of a and the value of b.

$\dfrac{(4 - \sqrt{18})}{\sqrt{2}} \times \dfrac{\sqrt{2}}{\sqrt{2}}$

First, rationalise the denominator.

$= \dfrac{4\sqrt{2} - \sqrt{36}}{(\sqrt{2})^2}$

$= \dfrac{4\sqrt{2} - 6}{2}$

Now simplify.

$= 2\sqrt{2} - 3$

Hence, $a = -3$ and $b = 2$

SUMMARY

- Surds are irrational numbers, which are any real numbers that cannot be expressed as a fraction $\dfrac{a}{b}$, where a and b are integers and $b \neq 0$.

- There are five rules when working with surds:

  $\sqrt{a} \times \sqrt{b} = \sqrt{ab}$

  $(\sqrt{b})^2 = \sqrt{b} \times \sqrt{b} = b$

  $\dfrac{\sqrt{a}}{\sqrt{b}} = \sqrt{\dfrac{a}{b}}$

  $(a + \sqrt{b})^2 = a^2 + 2a\sqrt{b} + b$

  $(a + \sqrt{b})(a - \sqrt{b}) = a^2 - b$

QUESTIONS

QUICK TEST

1. Express the following in the form $a\sqrt{b}$, simplifying the answers.

  a. $\sqrt{24}$

  b. $\sqrt{200}$

  c. $\sqrt{48} + \sqrt{12}$

2. Molly works out $(2 - \sqrt{3})^2$. She says the answer is 1. Decide whether Molly is correct, giving a reason for your answer.

3. Decide whether this is correct.
$\dfrac{1}{\sqrt{2}} = \dfrac{\sqrt{2}}{2}$

EXAM PRACTICE

1. Given that $\dfrac{5 - \sqrt{75}}{\sqrt{3}} = a + b\sqrt{3}$ find the value of a and the value of b.

2. Rationalise the denominator, simplifying the answer.
$\dfrac{3}{\sqrt{6}}$

3. Work out $\dfrac{(5 + \sqrt{5})(2 - 2\sqrt{5})}{\sqrt{45}}$

Give your answer in its simplest form.

Upper and Lower Bounds

Measurements

Measurements are never exact. They can only be expressed to a certain degree of accuracy.

When measurements are quoted to a given unit, say the nearest metre, there is a highest and lowest value they could be.

The highest value is called the **upper bound**.

The lowest value is called the **lower bound**.

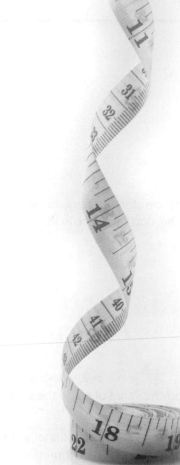

Example

6.2 cm rounded to the nearest millimetre would really lie between

$$6.15 \leqslant 6.2 < 6.25$$

6.15 is the lower bound

6.25 is the upper bound

> The real value can be as much as half the rounded unit below or above the rounded off value.

Finding Maximum and Minimum Possible Values of a Calculation

Some exam questions may ask you to calculate the upper or lower bounds of calculations.

	Upper bound	Lower bound
Addition	Upper bound + Upper bound	Lower bound + Lower bound
Multiplication	Upper bound × Upper bound	Lower bound × Lower bound
Subtraction	$\left(\begin{array}{c}\text{Upper bound of}\\\text{larger quantity}\end{array}\right) - \left(\begin{array}{c}\text{Lower bound of}\\\text{smaller quantity}\end{array}\right)$	$\left(\begin{array}{c}\text{Lower bound of}\\\text{larger quantity}\end{array}\right) - \left(\begin{array}{c}\text{Upper bound of}\\\text{smaller quantity}\end{array}\right)$
Division	$\dfrac{\text{Upper bound of quantity 1}}{\text{Lower bound of quantity 2}}$	$\dfrac{\text{Lower bound of quantity 1}}{\text{Upper bound of quantity 2}}$

Examples

1. A square measures 62 mm to the nearest mm. Work out the upper and lower bounds of the area of the square (in mm²).

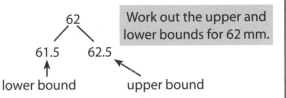

Work out the upper and lower bounds for 62 mm.

lower bound   upper bound

Upper bound $= 62.5 \times 62.5$

$\qquad\qquad = \textbf{3906.25 mm}^2$

Lower bound $= 61.5 \times 61.5$

$\qquad\qquad = \textbf{3782.25 mm}^2$

2. Given that $a = 4.2$ (to 1dp) and $b = 6.23$ (to 3sf), find the upper and lower bounds for the following calculations:

a. $\quad b - a$

Upper bound $= 6.235 - 4.15 = \textbf{2.085}$

Lower bound $= 6.225 - 4.25 = \textbf{1.975}$

b. $\quad \dfrac{b}{a}$

Upper bound $= \dfrac{6.235}{4.15} = \textbf{1.5024}$

Lower bound $= \dfrac{6.225}{4.25} = \textbf{1.4647}$

- Measurements are expressed to a certain degree of accuracy.
- The highest value is called the upper bound.
- The lowest value is called the lower bound.

QUESTIONS

QUICK TEST

1. $c = \dfrac{(2.6)^3 \times 12.52}{3.2}$

2.6 and 3.2 are correct to 1 decimal place. 12.52 is correct to 2 decimal places. Which of the following calculations gives the lower bound for c and which gives the upper bound for c?

A $\quad \dfrac{(2.65)^3 \times 12.525}{3.15}$

B $\quad \dfrac{(2.55)^3 \times 12.515}{3.15}$

C $\quad \dfrac{(2.65)^3 \times 12.525}{3.25}$

D $\quad \dfrac{(2.65)^3 \times 12.515}{3.15}$

E $\quad \dfrac{(2.55)^3 \times 12.515}{3.25}$

2. [123] Work out the upper and lower bound of $\dfrac{6}{p}$, where p is 27 (rounded to the nearest whole number).

EXAM PRACTICE

1. [123] The value of R is calculated by using this formula: $R = \dfrac{a - b}{b}$

$a = 7.65$ correct to 2 decimal places.

$b = 4.3$ correct to 1 decimal place.

Find the difference between the lower bound of R and the upper bound of R. Give your answer to 3 significant figures.

Formulae and Expressions

$a + b$ is called an **expression**. $b = a + 6$ is called a **formula**. The value of b depends on the value of a.

- A term is a collection of numbers, letters and brackets, all multiplied together, e.g. $6a$, $2ab$, $3(x-1)$.

- Terms are separated by + and – signs. Each term has a + or – sign in front of it.

$3ab$ means $3 \times a \times b$

$3b^2$ means $3 \times b \times b$

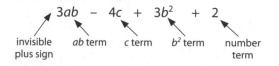

$$3ab \quad - \quad 4c \quad + \quad 3b^2 \quad + \quad 2$$

invisible ab term c term b^2 term   number
plus sign   term

Substituting into Formulae

Replacing a letter with a number is called **substitution**.

- Write out the expression first and then replace the letters with the values given.

- Work out the value, but take care with the order of operations, i.e. BIDMAS.

Examples

1. $a = 3b - 4c$

   Find a if $b = 4$ and $c = -2$

   $a = (3 \times 4) - (4 \times -2)$

   $= 12 - (-8)$

Taking away a negative is the same as adding.

   $= \mathbf{20}$

2. $E = \frac{1}{2}mv^2$

   Find E if $m = 6$ and $v = 10$

   $E = \frac{1}{2} \times 6 \times 10^2$

   $E = \mathbf{300}$

Rearranging Formulae

The subject of a formula is the letter that appears on its own on one side of the formula.

Examples

1. Make a the subject of the formula $b = (a - 3)^2$

   $$b = (a - 3)^2$$

   $$\pm\sqrt{b} = a - 3$$

Deal with the power first, square root both sides.

   $$\pm\sqrt{b} + 3 = a$$

Remove any term added or subtracted. Add 3 to both sides.

   $$a = \pm\sqrt{b} + 3 \text{ i.e. } \boldsymbol{a = 3 \pm \sqrt{b}}$$

2. Make x the subject of the formula $p = x^2 + y$

   $$p = x^2 + y$$

   $$p - y = x^2 \qquad \text{Subtract } y \text{ from both sides.}$$

   $$\pm\sqrt{p - y} = x \qquad \text{Square root both sides.}$$

   $$\boldsymbol{x = \pm\sqrt{p - y}}$$

3. Make t the subject of the formula $v = u + at$

   $$v = u + at$$

   $$v - u = at \qquad \text{Subtract } u \text{ from both sides.}$$

   $$\frac{\boldsymbol{v - u}}{\boldsymbol{a}} = \boldsymbol{t} \qquad \text{Divide all of } v - u \text{ by } a.$$

When rearranging a formula, sometimes the new subject appears in more than one term.

Example

Make x the subject of the formula $a = \dfrac{x+c}{x-d}$

$a = \dfrac{x+c}{x-d}$

$a(x-d) = x+c$ — Multiply both sides by $(x-d)$.

$ax - ad = x + c$ — Multiply out the brackets.

$ax - x = c + ad$ — Collect like terms involving x on one side of the equation.

$x(a-1) = c + ad$ — Factorise.

$x = \dfrac{c+ad}{a-1}$

These types of questions are usually worth at least 3 marks.

SUMMARY

- $a + b$ is an **expression**.
- $b = a + 6$ is a **formula**.
- The letter that appears on its own on one side of a formula is known as the **subject** of the formula.
- Letters multiplied together are usually placed in alphabetical order. The term ba is the same as the term ab.

QUESTIONS

QUICK TEST

1. Simplify the following expressions:

   a. $6a - 3b + 2a - 4b$

   b. $3a^2 - 6b^2 - 2b^2 + a^2$

   c. $5xy - 3yx + 2xy^2$

2. If $a = \dfrac{3}{5}$ and $b = -2$, find the value of these expressions:

   a. $ab - 5$

   b. $a^2 + b^2$

   c. $3a - 6ab$

3. Make u the subject of the formula

   $v^2 = u^2 + 2as$

4. Make p the subject of the formula

   $q = \dfrac{p-t}{p+v}$

EXAM PRACTICE

1. a. Sarah says 'when $x = 2$ the value of $3x^2$ is 36'.

      Josh says 'when $x = 2$ the value of $3x^2$ is 12'.

      Who is right? Explain why.

   b. Work out the value of $4(x-1)^2$ when $x = 4$.

   c. If $y = 4(x-1)^2$, make x the subject of the formula.

Brackets and Factorisation

Multiplying out brackets helps to simplify algebraic expressions.

Expanding Single Brackets

Each term outside the bracket multiplies each separate term inside the bracket.

$$5(x + 6) = 5x + 30$$

Examples

Expand and simplify:

1. $-2(2x + 4) = \mathbf{-4x - 8}$

2. $5(2x - 3) = \mathbf{10x - 15}$

3. $8(x + 3) + 2(x - 1)$ Multiply out the brackets.

$= 8x + 24 + 2x - 2$ Collect like terms.

$= \mathbf{10x + 22}$

4. $3(2x - 5) - 2(x - 3)$ Multiply out the brackets.

$= 6x - 15 - 2x + 6$ Collect like terms.

$= \mathbf{4x - 9}$

Expanding Two Brackets

Every term in the second bracket must be multiplied by every term in the first bracket.

Often, but not always, the two middle terms are like terms and can be collected together.

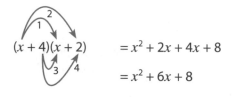

$$(x + 4)(x + 2) \qquad = x^2 + 2x + 4x + 8$$
$$= x^2 + 6x + 8$$

Examples

Expand and simplify:

1. $(x + 4)(2x - 5) = 2x^2 - 5x + 8x - 20$

$= \mathbf{2x^2 + 3x - 20}$

2. $(2x + 1)^2 = (2x + 1)(2x + 1)$

$= 4x^2 + 2x + 2x + 1$

$= \mathbf{4x^2 + 4x + 1}$

Remember that x^2 means x multiplied by itself.

3. $(3x - 1)(x - 2) = 3x^2 - 6x - x + 2$

$= \mathbf{3x^2 - 7x + 2}$

4. $(x - 4)(3x + 1) = 3x^2 + x - 12x - 4$

$= \mathbf{3x^2 - 11x - 4}$

5. $(2x + 3y)(x - 2y) = 2x^2 - 4xy + 3xy - 6y^2$

$= \mathbf{2x^2 - xy - 6y^2}$

$a \neq 0 \qquad f(x) = a(x$

Factorisation

Factorisation simply means putting an expression into brackets.

One Bracket

$4x + 6 = 2(2x + 3)$

To factorise $4x + 6$:

- Recognise that 2 is the Highest Common Factor of 4 and 6.

- Take out the common factor.

- The expression is completed inside the bracket so that when multiplied out it is equivalent to $4x + 6$.

Two Brackets

Two brackets are obtained when a quadratic expression of the type $ax^2 + bx + c$ is factorised.

Examples

1. $x^2 + 4x + 3 = (x + 1)(x + 3)$

2. $x^2 - 7x + 12 = (x - 3)(x - 4)$

3. $x^2 + 3x - 10 = (x + 5)(x - 2)$

4. $x^2 - 64 = (x - 8)(x + 8)$

This is known as the 'difference of two squares'. In general, $x^2 - a^2 = (x - a)(x + a)$.

5. $81x^2 - 25y^2 = (9x - 5y)(9x + 5y)$

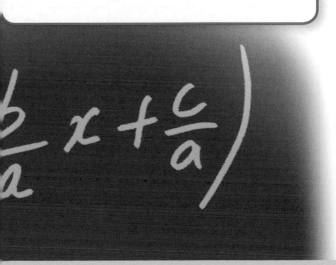

SUMMARY

- To expand single brackets, multiply each term outside the bracket by everything inside the bracket.

- To expand two brackets, multiply each term in the second bracket by each term in the first bracket.

- Factorising means putting in brackets. Look for the Highest Common Factor.

QUESTIONS

QUICK TEST

1. Expand and simplify:

   a. $(x + 3)(x - 2)$

   b. $4x(x - 3)$

   c. $(x - 3)^2$

2. Factorise:

   a. $12xy - 6x^2$

   b. $3a^2b + 6ab^2$

   c. $x^2 + 4x + 4$

   d. $x^2 - 4x - 5$

   e. $x^2 - 100$

EXAM PRACTICE

1. Expand and simplify:

   a. $t(3t - 4)$

   b. $4(2x - 1) - 2(x - 4)$

2. Factorise:

   a. $y^2 + y$ b. $5p^2q - 10pq^2$

   c. $(a + b)^2 + 4(a + b)$ d. $x^2 - 5x + 6$

Equations 1

Equations involve an unknown value that needs to be worked out.

Equations need to be kept balanced, so whatever is done to one side of the equation (for example, adding) also needs to be done to the other side.

Linear Equations of the Form $ax + b = c$

Examples

1. Solve: $3x = 15$

$x = \dfrac{15}{3}$ ← Divide both sides by 3.

$x = \mathbf{5}$

2. Solve: $\dfrac{x}{3} = 6$

$x = 6 \times 3$ ← Multiply both sides by 3.

$x = \mathbf{18}$

3. Solve: $5x - 2 = 13$

$5x = 13 + 2$ ← Add 2 to both sides.

$5x = 15$

$x = \dfrac{15}{5}$ ← Divide both sides by 5.

$x = \mathbf{3}$

4. Solve: $3x + 1 = 13$

$3x = 13 - 1$ ← Subtract 1 from both sides.

$3x = 12$

$x = \dfrac{12}{3}$ ← Divide both sides by 3.

$x = \mathbf{4}$

5. Solve: $\dfrac{x}{6} - 1 = 3$

$\dfrac{x}{6} = 3 + 1$ ← Add 1 to both sides.

$\dfrac{x}{6} = 4$

$x = 4 \times 6$ ← Multiply both sides by 6.

$x = \mathbf{24}$

Linear Equations of the Form $ax + b = cx +$

Examples

1. Solve:

$7x - 4 = 3x + 8$

$7x = 3x + 12$ ← Add 4 to both sides.

$4x = 12$

$x = \dfrac{12}{4}$ ← Subtract $3x$ from both sides.

$x = \mathbf{3}$

Check by substituting 3 into both sides of the equation:

$7 \times 3 - 4 = 17$

$3 \times 3 + 8 = 17$

Since both the left-hand side of the equation and the right-hand side of the equation give the same answer $x = 3$ is correct ✔

2. Solve:

$5x + 3 = 2x - 5$

$5x = 2x - 5 - 3$ ← Subtract 3 from both sides.

$5x = 2x - 8$

$5x - 2x = -8$ ← Subtract $2x$ from both sides.

$3x = -8$

$x = -\dfrac{8}{3}$

$x = \mathbf{-2\dfrac{2}{3}}$

Linear Equations with Brackets

Examples

1. Solve:

$$5(x - 1) = 3(x + 2)$$

$$5x - 5 = 3x + 6$$

$$5x = 3x + 11$$

$$2x = 11$$

$$x = \frac{11}{2}$$

$$x = \mathbf{5.5}$$

Just multiply out the brackets, then solve as normal.

2. Solve:

$$5(2x + 3) = 2(x - 6)$$

$$10x + 15 = 2x - 12$$

$$10x = 2x - 12 - 15$$

$$10x = 2x - 27$$

$$10x - 2x = -27$$

$$8x = -27$$

$$x = -\frac{27}{8}$$

$$x = \mathbf{-3\frac{3}{8}}$$

3. Solve:

$$\frac{3(2x - 1)}{5} = 6$$

$$3(2x - 1) = 6 \times 5$$

Multiply both sides by 5.

$$6x - 3 = 30$$

$$6x = 33$$

$$x = \frac{33}{6}$$

$$x = \mathbf{5.5}$$

SUMMARY

- Whatever you do to one side of an equation needs to be done to the other side of the equation as well.
- Work through the solution step by step.
- Check your solution by substituting in your answer.

QUESTIONS

QUICK TEST

Solve the following equations:

1. $2x - 6 = 10$

2. $5 - 3x = 20$

3. $4(2 - 2x) = 12$

4. $6x + 3 = 2x - 10$

5. $7x - 4 = 3x - 6$

6. $5(x + 1) = 3(2x - 4)$

EXAM PRACTICE

1. Solve the equations:

    a. $5x - 3 = 9$

    b. $7x + 4 = 3x - 6$

    c. $3(4y - 1) = 21$

2. Solve:

    a. $5 - 2x = 3(x + 2)$

    b. $\frac{3x - 1}{3} = 4 + 2x$

Equations 2

Equation Problems

When solving equation problems, the first step is to write down the information that you know.

Example

The perimeter of this rectangle is 30 cm.

Work out the value of y and find the length of the rectangle.

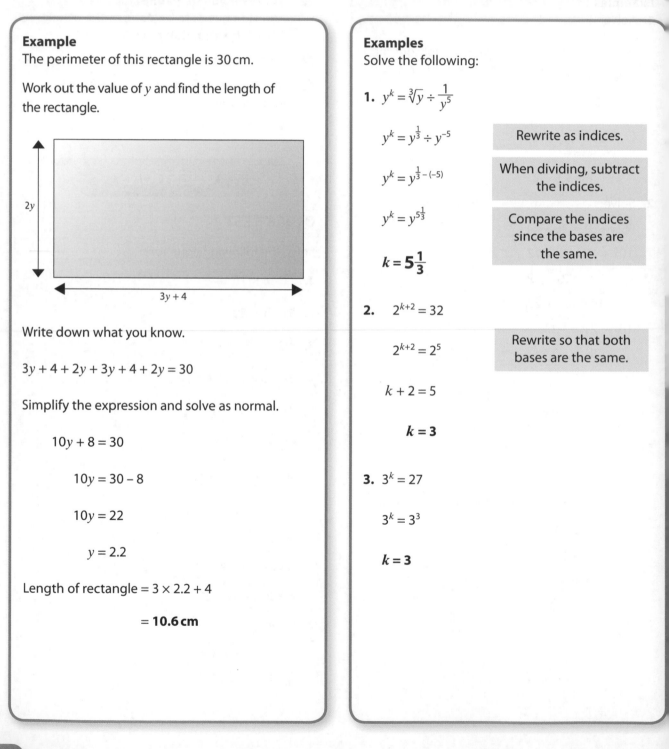

Write down what you know.

$$3y + 4 + 2y + 3y + 4 + 2y = 30$$

Simplify the expression and solve as normal.

$$10y + 8 = 30$$

$$10y = 30 - 8$$

$$10y = 22$$

$$y = 2.2$$

Length of rectangle $= 3 \times 2.2 + 4$

$$= \mathbf{10.6\,cm}$$

Solving Equations Involving Indices

Equations sometimes involve indices. You need to remember the laws of indices to be able to solve them

Examples

Solve the following:

1. $y^k = \sqrt[3]{y} \div \dfrac{1}{y^5}$

$y^k = y^{\frac{1}{3}} \div y^{-5}$ Rewrite as indices.

$y^k = y^{\frac{1}{3} - (-5)}$ When dividing, subtract the indices.

$y^k = y^{5\frac{1}{3}}$ Compare the indices since the bases are the same.

$$k = \mathbf{5\frac{1}{3}}$$

2. $2^{k+2} = 32$

$2^{k+2} = 2^5$ Rewrite so that both bases are the same.

$$k + 2 = 5$$

$$\mathbf{k = 3}$$

3. $3^k = 27$

$$3^k = 3^3$$

$$\mathbf{k = 3}$$

Example

The area of this rectangle is 81 cm².

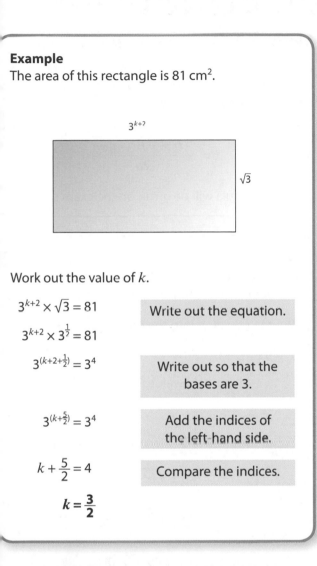

Work out the value of k.

$3^{k+2} \times \sqrt{3} = 81$ — Write out the equation.

$3^{k+2} \times 3^{\frac{1}{2}} = 81$

$3^{(k+2+\frac{1}{2})} = 3^4$ — Write out so that the bases are 3.

$3^{(k+\frac{5}{2})} = 3^4$ — Add the indices of the left-hand side.

$k + \frac{5}{2} = 4$ — Compare the indices.

$k = \dfrac{3}{2}$

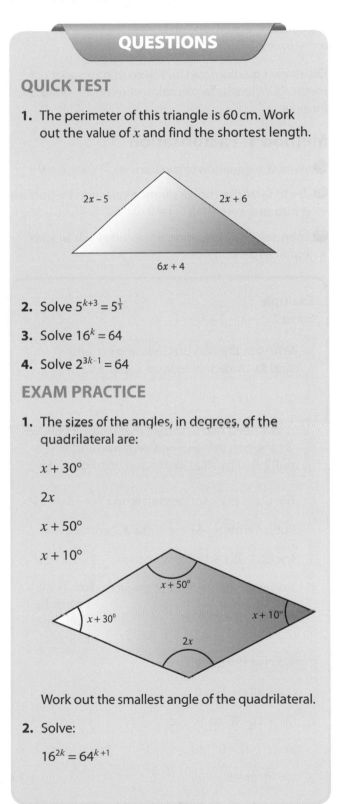

QUESTIONS

QUICK TEST

1. The perimeter of this triangle is 60 cm. Work out the value of x and find the shortest length.

2. Solve $5^{k+3} = 5^{\frac{1}{3}}$

3. Solve $16^k = 64$

4. Solve $2^{3k-1} = 64$

EXAM PRACTICE

1. The sizes of the angles, in degrees, of the quadrilateral are:

   $x + 30°$

   $2x$

   $x + 50°$

   $x + 10°$

   Work out the smallest angle of the quadrilateral.

2. Solve:

   $16^{2k} = 64^{k+1}$

Solving Quadratic Equations

A **quadratic equation** can be written in the form $ax^2 + bx + c = 0$.

Quadratic equations can be solved using a variety of methods including factorisation or using the quadratic formula.

Method 1: Factorisation

⬤ Write the equation in the form $ax^2 + bx + c = 0$

⬤ Try to factorise the quadratic expression by putting it into brackets $(\quad)(\quad) = 0$

⬤ Then solve the equation by putting each bracket equal to 0

Example
Solve $2x^2 - x - 3 = 0$

> Write out the two brackets and put an x and $2x$ in each one, since $x \times 2x = 2x^2$.

$(2x \quad)(x \quad)$

> We now need two numbers that multiply to give -3 (one positive and one negative) and when multiplied by the x terms add up to give $-1x$.

Try: $(2x + 1)(x - 3)$: Constant term $= 1 \times -3 = -3$ ✓

But: x terms $= -6x + x = -5x$ ✗ incorrect

Try: $(2x - 3)(x + 1)$: Constant term $= -3 \times 1 = -3$ ✓

x terms $= 2x - 3x = -x$ ✓ correct

> The $2x$ and 1 must be in different brackets.

A quick check gives
$(2x - 3)(x + 1) = 2x^2 - x - 3$

> Now solve the equation:

$(2x - 3)(x + 1) = 0$

$\therefore (2x - 3) = 0$ so $x = \dfrac{3}{2}$

or $(x + 1) = 0$ so $x = -1$

$\therefore x = \dfrac{3}{2}$ **or –1**

Method 2: The Quadratic Formula

When a quadratic expression does not factorise, use the quadratic formula. For any quadratic equation written in the form $ax^2 + bx + c = 0$,

$$x = \frac{-b \pm \sqrt{b^2 - 4ac}}{2a}$$

This formula will be given on the exam paper.

Example
Solve the equation $2x^2 - 7x = 5$
Give your answers to 2 decimal places.

> Put the equation into the form $ax^2 + bx + c = 0$.

$2x^2 - 7x - 5 = 0$

> Identify the values of a, b and c.

$a = 2, b = -7, c = -5$

> Substitute these values into the quadratic formula.

$$x = \frac{-b \pm \sqrt{b^2 - 4ac}}{2a}$$

$$x = \frac{7 \pm \sqrt{(-7)^2 - (4 \times 2 \times -5)}}{2 \times 2}$$

$$x = \frac{7 \pm \sqrt{49 - (-40)}}{4}$$

$$x = \frac{7 \pm \sqrt{89}}{4}$$

$x = \dfrac{7 + \sqrt{89}}{4}$

One solution is when we use $+\sqrt{89}$
$x = \mathbf{4.11}$ **(2dp)**

$x = \dfrac{7 - \sqrt{89}}{4}$

One solution is when we use $-\sqrt{89}$
$x = \mathbf{-0.61}$ **(2dp)**

Check
$2 \times (4.11)^2 - 7(4.11) - 5$
$= 0$ ✓

Check
$2 \times (-0.61)^2 - 7(-0.61) - 5$
$= 0$ ✓

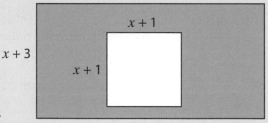

QUESTIONS

QUICK TEST

Solve:

1. a. $x^2 - 6x + 8 = 0$

   b. $x^2 + 5x + 4 = 0$

   c. $x^2 - 4x - 12 = 0$

2. Solve these equations by (i) factorisation (ii) using the quadratic formula.

   a. $x^2 + 2x - 15 = 0$

   b. $2x^2 + 5x + 2 = 0$

EXAM PRACTICE

1. Charles cuts a square out of a rectangular piece of card.

The length of the rectangle is $2x + 5$
The width of the rectangle is $x + 3$
The length of the side of the square is $x + 1$
All measurements are in centimetres

The shaded shape in the diagram shows the card remaining.

The area of the shaded shape is 45 cm².

a. Show that $x^2 + 9x - 31 = 0$

b. **i.** Solve the equation $x^2 + 9x - 31 = 0$
Give your answer correct to 3 significant figures.

     ii. Hence find the perimeter of the square.
Give your answer correct to 3 significant figures.

Solving Quadratic Equations and Cubic Equations

Completing the Square

Completing the square is when quadratic expressions, $ax^2 + bx + c$ are written in the form $(x+p)^2 + q$.

To complete the square of the expression $ax^2 + bx + c$

● If a is not 1, divide the whole expression by a

● Write the expression in the form $\left(x + \frac{b}{2}\right)^2$

Notice that the number in the bracket is always half the value of b, the coefficient of x

● Multiply out the brackets, compare to the original and adjust by adding and subtracting an extra amount

● Check : the value of $p = \frac{b}{2}$ and $q = -\left(\frac{b}{2}\right)^2 + c$

Example
The expression $x^2 + 8x + 7$ can be written in the form $(x+p)^2 + q$ for all values of x.

a. Find the values of p and q.

$x^2 + 8x + 7$ is in the form $ax^2 + bx + c$

$(x+4)^2$ | Half of 8 is 4

$(x+4)(x+4)$ $= x^2 + 8x + 16$ | Multiply out the brackets and now compare to the original $x^2 + 8x + 7$

$(x+4)^2 - 9$ | To make the expression equal we subtract 9

Hence **$p = 4$ and $q = -9$**

Check: $p = \frac{b}{2}$ $q = -\left(\frac{b}{2}\right)^2 + c$

$p = \frac{8}{2}$ $q = -\left(\frac{8}{2}\right)^2 + 7$

$p = 4$ $q = -(4)^2 + 7$

$q = -16 + 7$

$q = -9$

b. The expression $x^2 + 8x + 7$ has a minimum value. Find this minimum value.

Since $(x+4)^2 \geqslant 0$ for all values of x, the minimum value is equal to q.

Minimum value $= -9$

Solving Quadratic Equations by Completing the Square

● Rearrange the equation into the form $ax^2 + bx + c = 0$

● If a is not 1, divide every term by a

● Complete the square to get $(x+p)^2 + q = 0$

● Solve $(x+p)^2 = -q$

Example
Solve $x^2 = 6x - 2$ using the method of completing the square, giving your answers correct to 2 decimal places.

$x^2 = 6x - 2$

$x^2 - 6x + 2 = 0$ | Rearrange in the form $ax^2 + bx + c = 0$

$(x-3)^2$ | Half of -6 is -3

$(x-3)^2 = x^2 - 6x + 9$ | Multiply out the brackets and now compare to the original $x^2 - 6x + 2$

$(x-3)^2 - 7 = 0$ | To make the expression equal, we subtract 7
$(x-3)^2 - 7 = x^2 - 6x + 2$
$\therefore (x-3)^2 - 7 = 0$

Example (cont.)

Now solve:

$$(x - 3)^2 = 7$$

$(x - 3) = \pm\sqrt{7}$	Square root both sides. Remember ±.
$x = 3 \pm\sqrt{7}$	Add 3 to both sides.

$\therefore\quad x = 3 + \sqrt{7} = 5.645\ldots$ or $x = 3 - \sqrt{7} = 0.354\ldots$

$x = \mathbf{5.65}$ (2dp) **or** $x = \mathbf{0.35}$ (2dp)

Solving Cubic Equations by Trial and Improvement

Trial and improvement gives an approximate solution to cubic equations.

Example

The equation $x^3 + 2x = 58$ has a solution between 3 and 4. Find the solution to 1 decimal place.

Drawing a table can help you, and also help the examiner since it makes it easier to follow what you have done.

x	$x^3 + 2x$	Comment
3.5	$3.5^3 + 2 \times 3.5 = 49.875$	too small
3.8	$3.8^3 + 2 \times 3.8 = 62.472$	too big
3.7	$3.7^3 + 2 \times 3.7 = 58.053$	too big
3.6	$3.6^3 + 2 \times 3.6 = 53.856$	too small
3.65	$3.65^3 + 2 \times 3.65 = 55.927$	too small

So $x = \mathbf{3.7}$ (1dp)

Since the exact value of the x is between 3.65 and 3.7

QUESTIONS

QUICK TEST

1. 123 The equation $a^3 = 40 - a$ has a solution between 3 and 4. Find the solution to 1 decimal place, by using a method of trial and improvement.

2. The expression $x^2 - 4x + 7$ can be written in the form $(x + p)^2 + q$ for all values of x. Find the values of p and q.

EXAM PRACTICE

1. 123 The equation $x^3 + 4x^2 = 49$ has a solution between $x = 2$ and $x = 3$.

   Use a trial and improvement method to find this solution. Give your answer correct to 1 decimal place. You must show all your working.

2. The expression $x^2 + 10x + 5$ can be written in the form $(x + a)^2 + b$ for all values of x.

   a. Find a and b.

   b. The expression $x^2 + 10x + 5$ has a minimum value. Find this minimum value.

Simultaneous Linear Equations

Two equations with two unknowns are called **simultaneous equations**. They can be solved algebraically or graphically.

Solving Algebraically (Elimination Method)

Solve simultaneously:	$3x + 2y = 8$ $2x - 3y = 14$

Label the equations ① and ②.	$3x + 2y = 8$ ① $2x - 3y = 14$ ②

Since no coefficients match, multiply equation ① by 2 and equation ② by 3.	$6x + 4y = 16$ $6x - 9y = 42$

Rename them equations ③ and ④.	$6x + 4y = 16$ ③ $6x - 9y = 42$ ④

The coefficient of x in equations ③ and ④ is the same. Subtract equation ④ from equation ③ and solve to find y.	$0x + 13y = -26$ $y = -26 \div 13$ $y = -2$	Note $4y - (-9y)$ $= 4y + 9y$ $= 13y$

Substitute the value of $y = -2$ into equation ①. Solve this equation to find x.	$3x + 2 \times (-2) = 8$ $3x + (-4) = 8$ $3x = 8 + 4$ $3x = 12$ $x = 4$	You could substitute into ②.

Check in equation ②.	$(2 \times 4) - (3 \times -2) = 14$ ✔

Solution is: $x = 4$, $y = -2$

Solving Graphically

The point at which any two graphs intersect represents the simultaneous solutions of their equations.

Example

Solve the simultaneous equations:

$2x + 3y = 6$

$x + y = 1$

Draw the graph of:

$2x + 3y = 6$

When $x = 0$, $3y = 6$ ∴ $y = 2$ $(0, 2)$

When $y = 0$, $2x = 6$ ∴ $x = 3$ $(3, 0)$

Draw the graph of:

$x + y = 1$

When $x = 0$, $y = 1$ $(0, 1)$

When $y = 0$, $x = 1$ $(1, 0)$

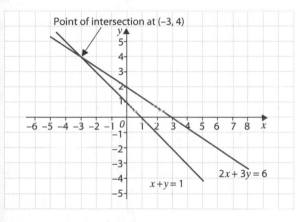

Point of intersection at $(-3, 4)$

At the point of intersection: $x = -3$, $y = 4$. This is the solution of the simultaneous equations.

QUICK TEST

1. Solve the simultaneous equations:

   $4b + 7a = 10$

   $2b + 3a = 3$

2. The diagram shows the graphs of the lines:

   $x + y = 6$ and $y = x + 2$

   Use the diagram to solve the simultaneous equations $x + y = 6$ and $y = x + 2$.

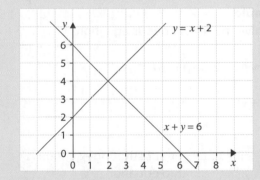

EXAM PRACTICE

1. Solve the simultaneous equations:

   $5a - 2b = 19$

   $3a + 4b = 1$

2. Frances and Patrick were organising a children's party. They went to the toy shop and bought some hats and balloons. Frances bought 5 hats and 4 balloons. She paid £22. Patrick bought 3 hats and 5 balloons. He paid £21.

   The cost of a hat was x pounds. The cost of a balloon was y pounds. Work out the cost of one hat and the cost of one balloon.

Solving a Linear and a Non-linear Equation Simultaneously

You need to be able to work out the coordinates of the points of intersection of a straight line and a quadratic curve, as well as a straight line and a circle.

Point of Intersection of a Straight Line and a Quadratic Graph

Example

a. Solve the simultaneous equations:

$y = 4x - 2$ ①

$y = x^2 + 1$ ②

> Eliminate y by substituting equation ② into equation ①.

$$x^2 + 1 = 4x - 2$$

Rearrange: $x^2 - 4x + 3 = 0$

Factorise: $(x - 1)(x - 3) = 0$

Solve: $x = 1$ and $x = 3$

> Now substitute the values of x back into equation ② to find the corresponding values of y.

When $x = 1$, $y = 1^2 + 1 = 2$; $\boldsymbol{x = 1, y = 2}$

When $x = 3$, $y = 3^2 + 1 = 10$; $\boldsymbol{x = 3, y = 10}$

b. Give a geometrical interpretation of the result.

The diagram shows the points of intersection A and B of the line $y = 4x - 2$ and the curve $y = x^2 + 1$.

We know that the points A and B lie on both the curve and the line.

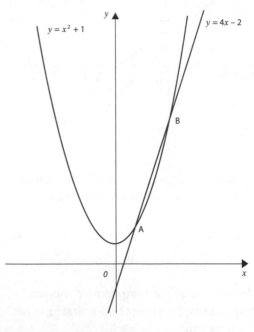

Coordinates of A are (1, 2).

Coordinates of B are (3, 10).

Points of Intersection of a Straight Line and a Circle

The equation of a circle with centre (0, 0) and radius r is $x^2 + y^2 = r^2$.

Example

Find the coordinates of the points where the line $y - x = 6$ cuts $x^2 + y^2 = 36$.

The coordinates must satisfy both equations, so we solve them simultaneously.

$y - x = 6$ ①
$x^2 + y^2 = 36$ ②

> Rewrite equation ① in the form '$y =$'.

$y = x + 6$ ①

> Eliminate y by substituting equation ① into equation ②.

$x^2 + (x + 6)^2 = 36$

> Expand brackets: $(x + 6)^2 = (x + 6)(x + 6)$
> $\qquad\qquad\qquad = x^2 + 12x + 36$

$x^2 + x^2 + 12x + 36 = 36$
Rearrange: $2x^2 + 12x = 0$

Factorise: $2x(x + 6) = 0$

Solve: $x = 0, x = -6$

> Substitute the values of x into equation ① to find the values of y.

Since $y - x = 6$, $y = x + 6$

When $x = 0$, $y = 0 + 6 = 6$

When $x = -6$, $y = -6 + 6 = 0$

The line $y - x = 6$ cuts the circle $x^2 + y^2 = 36$ at (0, 6) and (–6, 0).

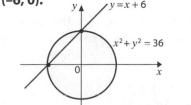

SUMMARY

- Two equations with two unknowns are called simultaneous equations.

- Simultaneous equations can be solved by a graphical method or by a method of elimination or substitution.

- The equation $x^2 + y^2 = r^2$ is an equation of a circle with centre (0, 0) and radius r.

QUESTIONS

QUICK TEST

1. Solve these simultaneous equations:

   a. $y = x + 4$

   $y - x^2 + 2$

   b. $y = x + 1$

   $x^2 + y^2 = 25$

2. For each of the two questions above, give a geometrical interpretation of the result.

EXAM PRACTICE

1. Find the coordinates of the points where the line $y - 5 = x$ cuts the circle $x^2 + y^2 = 25$.

2. Solve the simultaneous equations:

   $x^2 + y^2 = 26$

   $y = 2x + 3$

Algebraic Fractions

Simplifying Algebraic Fractions

When working with algebraic fractions, the same rules are used as with ordinary fractions.

> **Example**
>
> Simplify $\dfrac{8x + 16}{x^2 - 4}$
>
> First, you need to factorise the numerator and denominator.
>
> $\dfrac{8(x + 2)}{(x + 2)(x - 2)}$
>
> **Remember the difference of two squares:**
> $a^2 - b^2 = (a + b)(a - b)$
>
> Now cancel any common factors, i.e. the $(x + 2)$.
>
> $\dfrac{8\cancel{(x + 2)}^1}{\cancel{(x + 2)}_1 (x - 2)} = \dfrac{8}{x - 2}$

Multiplication

When multiplying algebraic fractions, multiply the numerators and multiply the denominators and then cancel if possible.

> **Example**
>
> Solve $\dfrac{3a^2}{4b} \times \dfrac{16b^2}{9ab}$
>
>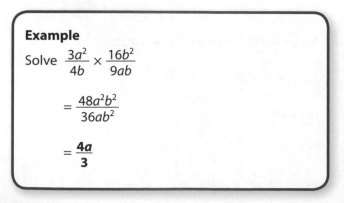
>
> $= \dfrac{48a^2b^2}{36ab^2}$
>
> $= \dfrac{4a}{3}$

Division

When dividing algebraic fractions, remember to turn the second fraction upside down (i.e. take the reciprocal) and then multiply and cancel if possible.

> **Example**
>
> Solve $\dfrac{12(x - 3)}{(x + 2)} \div \dfrac{4(x - 1)(x - 3)}{(x + 1)}$
>
> Cancel out common factors.
>
>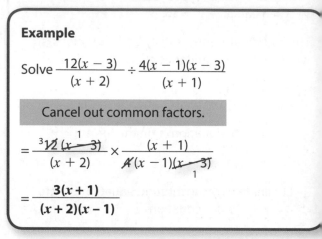
>
> $= \dfrac{3(x + 1)}{(x + 2)(x - 1)}$

Addition and Subtraction

As with ordinary fractions, you cannot add or subtract unless the fractions have the same denominator.

> **Example**
>
> Solve $\dfrac{(x + 3)}{(x + 2)} + \dfrac{3}{(x - 1)}$
>
> The common denominator is:
> $(x + 2)(x - 1)$.
>
> The numerator now needs to be adjusted.
>
> $\dfrac{(x + 3)(x - 1) + 3(x + 2)}{(x + 2)(x - 1)}$
>
> Multiply out the numerator.
>
> $\dfrac{x^2 + 2x - 3 + 3x + 6}{(x + 2)(x - 1)}$
>
> Simplify, then check whether the numerator can by factorised and the expression simplified further by cancelling.
>
> $= \dfrac{x^2 + 5x + 3}{(x + 2)(x - 1)}$

Solving Equations with Algebraic Fractions

Equations with algebraic fractions sometimes lead to quadratic equations, which can then be solved as shown previously.

Example

Solve $\dfrac{3}{x+1} + \dfrac{4}{x+2} = 2$

> The common denominator is $(x+1)(x+2)$. Now adjust the numerator.

$$\dfrac{3(x+2) + 4(x+1)}{(x+1)(x+2)} = 2$$

> Expand the brackets and simplify.

$$\dfrac{3x+6+4x+4}{x^2+3x+2} = 2$$

> Multiply the right-hand side by $x^2 + 3x + 2$

$$7x + 10 = 2(x^2 + 3x + 2)$$

$$7x + 10 = 2x^2 + 6x + 4$$

> Make the equation equal to zero.

$$2x^2 - x - 6 = 0$$

> Factorise, then solve the quadratic equation.

$$(2x + 3)(x - 2) = 0$$

So either $x = -\dfrac{3}{2}$ or $x = 2$

SUMMARY

● With algebraic fractions, the same rules are applied as with ordinary fractions.

- When adding or subtracting, make sure that the denominators are the same.

- When multiplying, multiply the numerators and multiply the denominators.

- When dividing, turn the second fraction upside down (take the reciprocal) and then multiply.

QUESTIONS

QUICK TEST

1. Decide whether these expressions are fully simplified:

   a. $\dfrac{6s^2w}{3w^2}$ b. $\dfrac{5a^2b}{2c}$ c. $\dfrac{9abc}{d}$

2. Simplify the following algebraic fractions:

   a. $\dfrac{3}{(x+1)} + \dfrac{2}{(x-1)}$

   b. $\dfrac{3a^2b}{2q} \times \dfrac{4q^2}{9abc}$

   c. $\dfrac{10(a^2 - b^2)}{3x + 6} \div \dfrac{(a-b)}{4x+8}$

EXAM PRACTICE

1. Simplify fully: $\dfrac{8x + 16}{x^2 - 4}$

2. Simplify fully: $\dfrac{x^2(6 + x)}{x^2 - 36}$

3. Simplify fully: $\dfrac{2x^2 + 7x - 15}{x^2 + 3x - 10}$

4. Solve: $\dfrac{3}{x} - \dfrac{4}{x+1} = 1$

Inequalities

In **inequalities** the left-hand side does not equal the right-hand side.

Inequalities can be solved in exactly the same way as equations, except that when multiplying or dividing by a negative number you must reverse the inequality sign.

The Inequality Symbols

$>$ means **greater than**

$<$ means **less than**

$\geqslant$ means **greater than or equal to**

$\leqslant$ means **less than or equal to**

Examples

1. Solve:

$$2x - 2 < 10$$
$$2x < 10 + 2$$
$$2x < 12$$
$$\mathbf{x < 6}$$

2. Solve:

$$3 - 2x \geqslant 9$$
$$-2x \geqslant 9 - 3$$
$$-2x \geqslant 6$$
$$x \leqslant \frac{6}{-2}$$

Divide by −2 and reverse the inequality.

$$\mathbf{x \leqslant -3}$$

3. Solve:

$$-7 < 3x - 1 \leqslant 11$$

Add 1 to each part of the inequality.

$$-6 < 3x \leqslant 12$$

Divide each part of the inequality by 3.

$$\mathbf{-2 < x \leqslant 4}$$

The integer values that satisfy this inequality are −1, 0, 1, 2, 3, 4.

Number Lines

Inequalities can be shown on a number line.

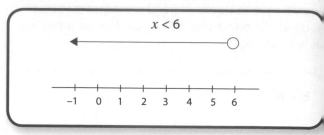

$x < 6$

The open circle means that 6 is not included.

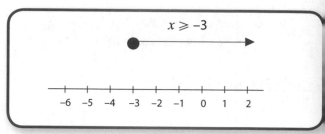

$x \geqslant -3$

The solid circle means that −3 is included.

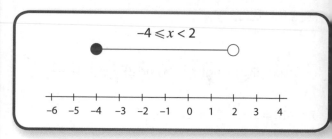

$-4 \leqslant x < 2$

The integer values that satisfy this inequality are −4, −3, −2, −1, 0, 1.

Graphs of Inequalities

The graph of an equation such as $x = 2$ is a line, whereas the graph of the inequality $x < 2$ is a region that has $x = 2$ as its boundary.

The diagram below shows unshaded the region R:

$x + y \leqslant 7$

$x > 2$

$y \geqslant 2$

For strict inequalities $>$ and $<$ the boundary line is not included and is shown as a dashed line.

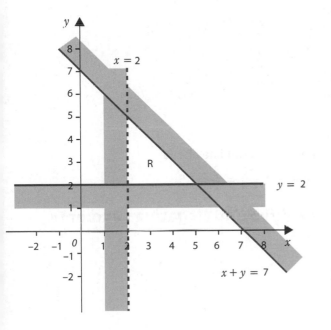

SUMMARY

- $>$ **greater than**

  $<$ **less than**

  $\geqslant$ **greater than or equal to**

  $\leqslant$ **less than or equal to**

- **On number lines, an open circle means the value is not included in the inequality and a solid circle means the value is included in the inequality.**

- **On graphs of inequalities, use a dashed line when the boundary is not included.**

QUESTIONS

QUICK TEST

1. Solve the following inequalities:

   a. $5x - 1 < 10$

   b. $6 \leqslant 3x + 2 < 11$

   c. $3 - 5x < 12$

EXAM PRACTICE

1. **a.** n is an integer such that $-6 < 2n \leqslant 8$.

   List all the possible values of n.

   b. Solve the inequality $4 + x > 7x - 8$.

2. On the diagram below, leave unshaded the region satisfied by these inequalities:

   $x + y \leqslant 5$
   $x \geqslant 1$
   $y > 1$

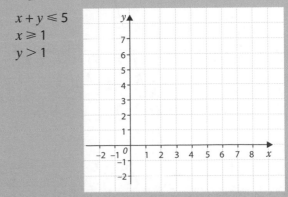

Straight-line Graphs

Drawing Straight-line Graphs

The general equation of a straight line is:

> $$y = mx + c$$
> m is the gradient. c is the intercept on the y-axis.

To draw the graph of $y = 3x - 4$:

⚫ Work out the coordinates of the points that lie on the line $y = 3x - 4$, by drawing a table of values of x. Substitute the x values into the equation $y = 3x - 4$, to find the values of y.
e.g. $x = 2, y = 3 \times 2 - 4 = 2$

x	−1	0	2	4
y	−7	−4	2	8

⚫ The coordinates of the points on the line are:

(−1, −7)   (0, −4)   (2, 2)   (4, 8)

Just read them from the table of values.

⚫ Plot the points (across the whole grid) and join with a straight line.

⚫ The line $y = 3x - 4$ is drawn.

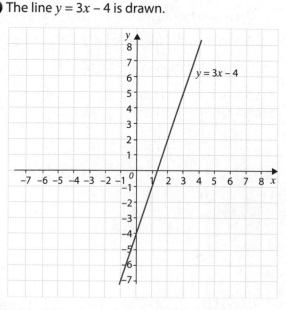

⚫ Label the line once you have drawn it.

Gradient of a Straight Line

Be careful when finding the gradient: double-check the scales.

$$\text{Gradient} = \frac{\text{change in } y}{\text{change in } x}$$

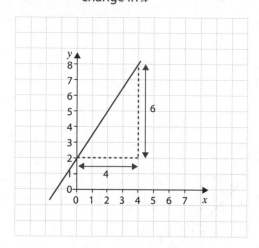

For the straight line above:

$$\text{Gradient} = \frac{6}{4} = \frac{3}{2} = 1.5$$

Positive and Negative Gradients

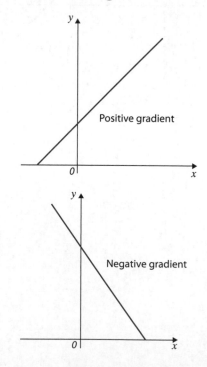

Finding the Midpoint of a Line Segment

The midpoint of a line segment between two points can be found by finding the mean of the x coordinates and the mean of the y coordinates of the points.

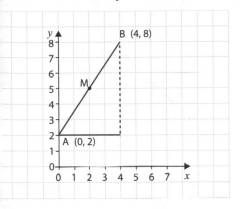

The midpoint of a line that joins the point $A(x_1, y_1)$ and $B(x_2, y_2)$ is:

$$\left(\frac{(x_1 + x_2)}{2}, \frac{(y_1 + y_2)}{2} \right)$$

The midpoint, M, of the line AB drawn here is:

$$\left(\frac{(0 + 4)}{2}, \frac{(2 + 8)}{2} \right) = (2, 5)$$

Perpendicular Lines

If two lines are **perpendicular** the product of their gradients is –1.

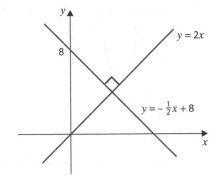

The lines $y = 2x$ and $y = -\frac{1}{2}x + 8$ are perpendicular because the gradients multiply to give –1.

$$2 \times -\frac{1}{2} = -1)$$

QUESTIONS

QUICK TEST

1. **a.** Complete the table of values for $y = 2x + 3$.

x	–2	–1	0	1	2	3
y						

   b. Draw the graph of $y = 2x + 3$.

EXAM PRACTICE

1.

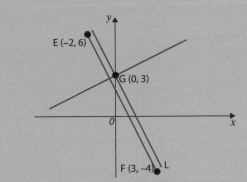

The diagram shows three points: E (–2, 6), F (3, –4) and G (0, 3). A line L is parallel to EF and passes through G.

   a. Find an equation for the line L.

   b. Find the midpoint of the line EF.

   c. Find the equation of a line perpendicular to line L and which passes through point G.

Curved Graphs

Quadratic Graphs

Quadratic graphs are of the form $y = ax^2 + bx + c$ where $a \neq 0$. Quadratic graphs have an x^2 term as the highest power of x.

They will be $\cup$ shaped if the coefficient of x^2 is positive, and $\cap$ shaped if the coefficient of x^2 is negative.

To draw the graph of $y = x^2 - 2x - 6$ using values of x from −2 to 4:

● Draw a table of values.

  Fill in the table of values by substituting the values of x into the equation.

  e.g. $x = 1, \quad y = 1^2 - 2 \times 1 - 6 = -7$
       Coordinates are (1, −7)

x	−2	−1	0	1	2	3	4
y	2	−3	−6	−7	−6	−3	2

● Draw the axes on graph paper and plot the points.

● Join the points with a smooth curve.

● Label the curve.

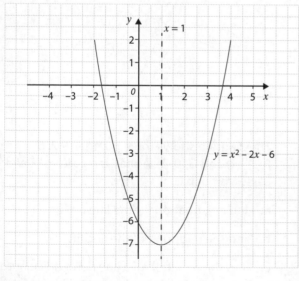

The minimum point is (1, −7).

The line of symmetry is $x = 1$.

Graph Shapes

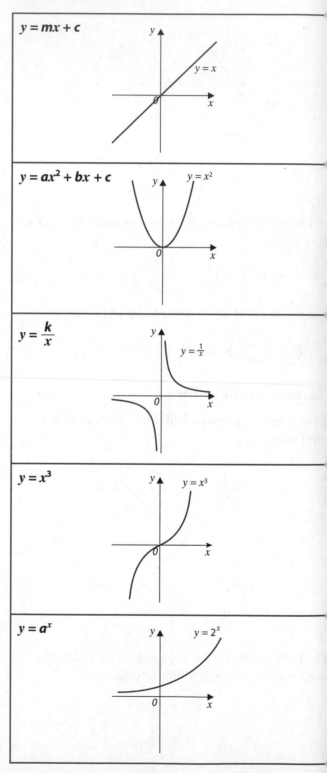

$y = mx + c$

$y = ax^2 + bx + c$

$y = \dfrac{k}{x}$

$y = x^3$

$y = a^x$

Example
Match each graph to one of the following equations.

$y = x^2 - 4$ $y = 5 - 2x$ $y = x^3$ $y = 3x - 1$

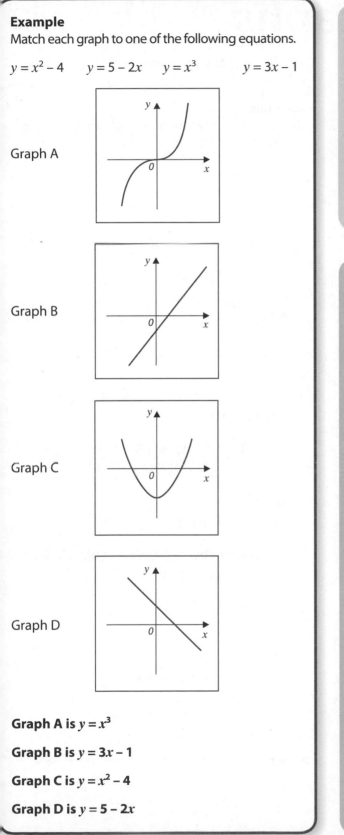

Graph A

Graph B

Graph C

Graph D

Graph A is $y = x^3$

Graph B is $y = 3x - 1$

Graph C is $y = x^2 - 4$

Graph D is $y = 5 - 2x$

SUMMARY

- Make sure you know the different graph shapes.
- Quadratic graphs have an x^2 term as the highest power of x. They are $\cup$ shaped or $\cap$ shaped.
- Cubic graphs have an x^3 term as the highest power of x.
- The reciprocal function is $y = \dfrac{k}{x}$

QUESTIONS

QUICK TEST

1. Match each graph below to one of the equations.

   $y = x^3 - 5$ $y = 2 - x^2$

   $y = 4x + 2$ $y = \dfrac{3}{x}$

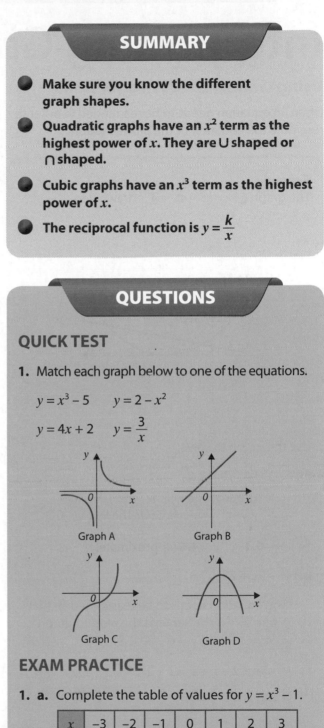

Graph A   Graph B

Graph C   Graph D

EXAM PRACTICE

1. **a.** Complete the table of values for $y = x^3 - 1$.

x	−3	−2	−1	0	1	2	3
y				−1			

   b. Using a suitable scale, draw the graph of $y = x^3 - 1$.

   c. From the graph find the approximate value of x when $y = 15$.

Interpreting Graphs

Using Graphs to Solve Equations

Often an equation needs to be rearranged in order to resemble the equation of the plotted graph.

Example
The graph of $y = x^2 - 4x + 4$ is drawn here:

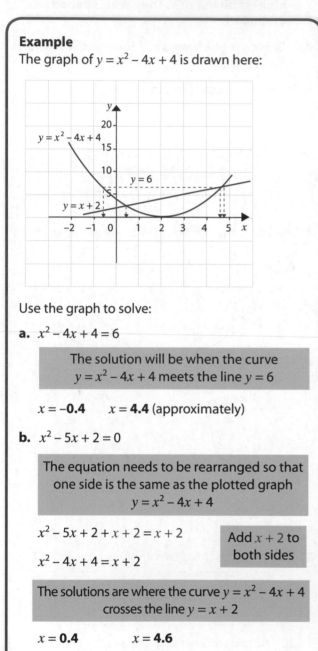

Use the graph to solve:

a. $x^2 - 4x + 4 = 6$

> The solution will be when the curve $y = x^2 - 4x + 4$ meets the line $y = 6$

$x = -0.4$ $x = 4.4$ (approximately)

b. $x^2 - 5x + 2 = 0$

> The equation needs to be rearranged so that one side is the same as the plotted graph $y = x^2 - 4x + 4$

$x^2 - 5x + 2 + x + 2 = x + 2$ Add $x + 2$ to both sides

$x^2 - 4x + 4 = x + 2$

> The solutions are where the curve $y = x^2 - 4x + 4$ crosses the line $y = x + 2$

$x = 0.4$ $x = 4.6$

Using Graphs to Find Relationships

Example
This graph is known to fit $y = pq^x$, where p and q are positive constants.

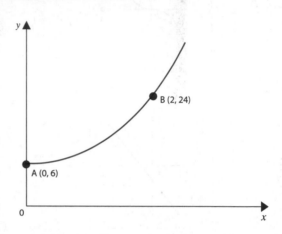

Use the graph to find the values of p and q and the relationship.

> Substitute the coordinates of point A into $y = pq^x$

$6 = p \times q^0$
Since $q^0 = 1$ **$p = 6$**

> Substitute the coordinates of point B into $y = pq^x$

$24 = 6 \times q^2$
$\therefore$ $4 = q^2$
  $q = 2$

The relationship is $y = 6 \times 2^x$

Speed–Time Graphs

A speed–time graph is useful when finding the acceleration or deceleration of an object. It can also be used to find the distance travelled.

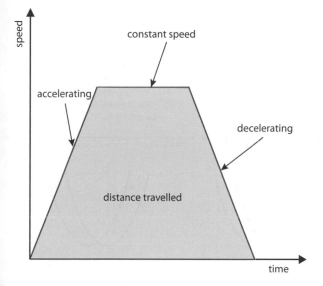

- Distance is the area between the graph and x-axis.
- A positive gradient means the speed is increasing.
- A negative gradient means the speed is decreasing.
- A horizontal line means the speed is constant.

SUMMARY

- **Accurate graphs can be used to find solutions to some equations.**

- **Speed–time graphs can be used to find the distance travelled by an object and also acceleration and deceleration.**

QUESTIONS

QUICK TEST

Draw the graph of $y = x^2 + x - 2$.
Use your graph to decide whether the solutions of these equations are true or false.

1. $x^2 + x = 5$
   solution is approximately $x = -2.8$, $x = 1.8$

2. $x^2 - 2 = 0$
   solution is approximately $x = 2.4$, $x = -2.4$

3. $x^2 - 2x - 2 = 0$
   solution is approximately $x = -0.7$, $x = 5.6$

EXAM PRACTICE

1. Miss Jones has a car.

   The value of the car on 1 September 2008 was £9000. The value of the car on 1 September 2010 was £1000.

   The sketch graph shows how the value, £v, of the car changes with time. The equation of the curve is:

   $v = ab^t$

   where t is the number of years after 1 September 2008, and a and b are positive constants.

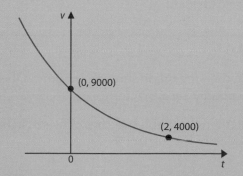

   Use the information on the graph to find the values of a and b.

Functions and Transformations

If y is equal to an expression involving x, then it can be written as $y = f(x)$. The graphs of the related functions can be found by applying transformations.

$y = f(x) \pm a$

This is when the graphs move up or down the y-axis by a value of a, i.e. a translation of $\begin{pmatrix} 0 \\ \pm a \end{pmatrix}$

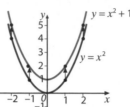

Graph moves up the y-axis by one unit.

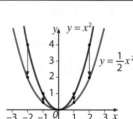

Graph moves down the y-axis by one unit.

$y = f(x \pm a)$

This is when the graphs move along the x-axis by a units.

$y = f(x + a)$ moves the graph a units to the left, i.e. a translation of $\begin{pmatrix} -a \\ 0 \end{pmatrix}$

$y = f(x - a)$ moves the graph a units to the right, i.e. a translation of $\begin{pmatrix} a \\ 0 \end{pmatrix}$

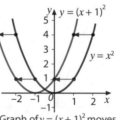

Graph of $y = (x + 1)^2$ moves one unit to the left.

Graph of $y = (x - 1)^2$ moves one unit to the right.

$y = kf(x)$

This is when the original graph stretches along the y-axis by a factor of k.

If $k > 1$, then the points are stretched upwards in the y direction of a scale factor of k.

If $k < 1$, e.g. $y = \frac{1}{2}x^2$, the graph squashes downwards by a scale factor of $\frac{1}{2}$.

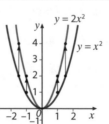

For this graph the x values stay the same and the y values are multiplied by 2.

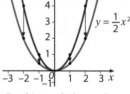

For this graph the x values stay the same and the y values are multiplied by $\frac{1}{2}$.

$y = f(kx)$

If $k > 1$, then the graph stretches inwards in the x direction by a scale factor $\frac{1}{k}$.

If $k < 1$, then the graph stretches outwards in the x direction by a scale factor $\frac{1}{k}$.

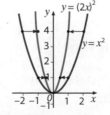

For this graph the y coordinates stay the same and the x values are multiplied by $\frac{1}{2}$.

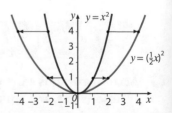

For this graph the y coordinates stay the same and the x values are multiplied by 2.

Applying Reflections

$= -f(x)$ This is a reflection in the x-axis.

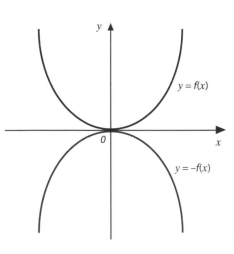

$= f(-x)$ This is a reflection in the y-axis.

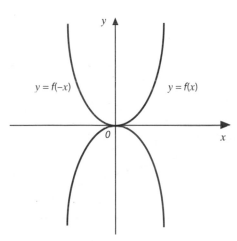

QUESTIONS

QUICK TEST

1. The graph of $y = f(x)$ is shown on the grid.

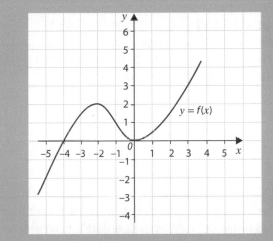

   a. Sketch the graph of $y = f(x) + 1$.

   b. Sketch the graph of $y = -f(x)$.

EXAM PRACTICE

1. This is a sketch of the curve $y = f(x)$.

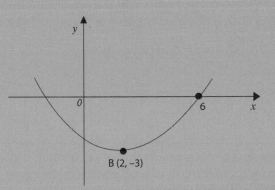

   The minimum point of the curve is point B (2, −3).

   Write down the coordinates of the minimum point for each of the following curves:

   a. $y = f(x + 3)$ **b.** $y = f(x) - 4$

   c. $y = f(x - 2)$ **d.** $y = f(-x)$

   e. $y = f(x + 1)$

Loci

The **locus** of a point is the set of all the possible positions that the point can occupy, subject to some given condition or rule.

Types of Loci

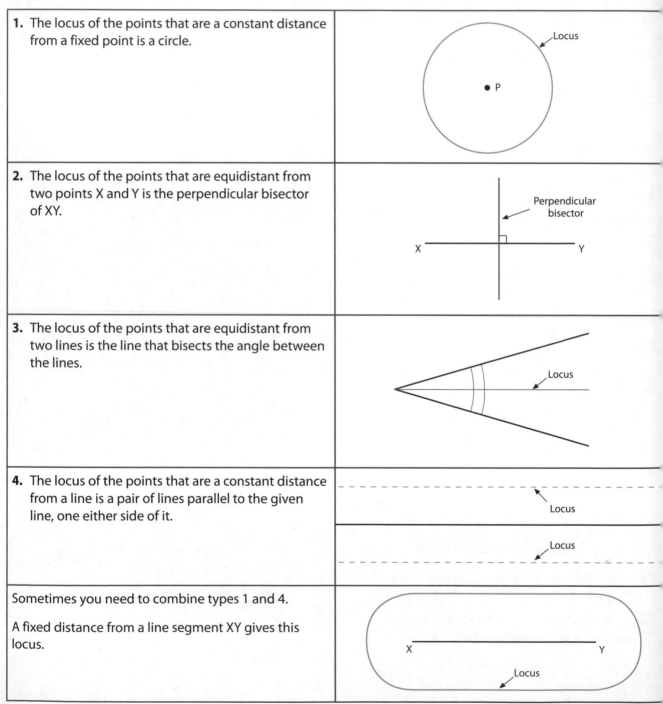

1. The locus of the points that are a constant distance from a fixed point is a circle.	
2. The locus of the points that are equidistant from two points X and Y is the perpendicular bisector of XY.	
3. The locus of the points that are equidistant from two lines is the line that bisects the angle between the lines.	
4. The locus of the points that are a constant distance from a line is a pair of lines parallel to the given line, one either side of it.	
Sometimes you need to combine types 1 and 4. A fixed distance from a line segment XY gives this locus.	

Example

Three radio transmitters form an equilateral triangle ABC with sides of 50 km. The range of the transmitter at A is 37.5 km, at B 30 km and at C 28 km. Using a scale of 1 cm to 10 km, construct a scale diagram to show where signals from all three transmitters can be received.

Below is a sketch not drawn to scale.

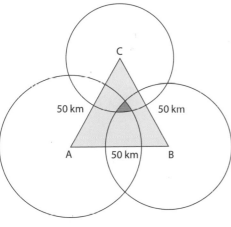

Please note that on your scale drawing the circle at A would have a radius of 3.75 cm. The circle at B would have a radius of 3 cm and the circle at C a radius of 2.8 cm.

The area where signals from all three transmitters can be received is shaded dark blue.

SUMMARY

- **Remember the four main types of loci:**

  - **A constant distance from a fixed point is a circle.**

  - **Equidistant from two fixed points is the perpendicular bisector of the line segment joining the points.**

  - **Equidistant from two lines is the bisector of the angle between the lines.**

  - **A constant distance from a line is a pair of parallel lines above and below.**

QUESTIONS

QUICK TEST

1. ABCD is a rectangle. The rectangle is accurately drawn.

Shade the set of points inside the rectangle which are more than 2 cm from point B and more than 1.5 cm from the line AD.

EXAM PRACTICE

1. The plan shows a garden drawn to a scale of 1 cm : 2 m. A and B are bushes and C is a pond.

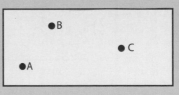

A landscape gardener has decided:

a. to lay a path right across the garden at an equal distance from each of the bushes, A and B.

b. to lay a flower border 2 m wide around pond C.

Construct these features on the plan above.

Translations and Reflections

Translations

Translations move figures from one position to another position. **Vectors** are used to describe the distance and direction of the translations.

A vector is written as $\begin{pmatrix} a \\ b \end{pmatrix}$.

a represents the horizontal distance and b represents the vertical distance.

The object and the image are **congruent** when the shape is translated, i.e they are identical.

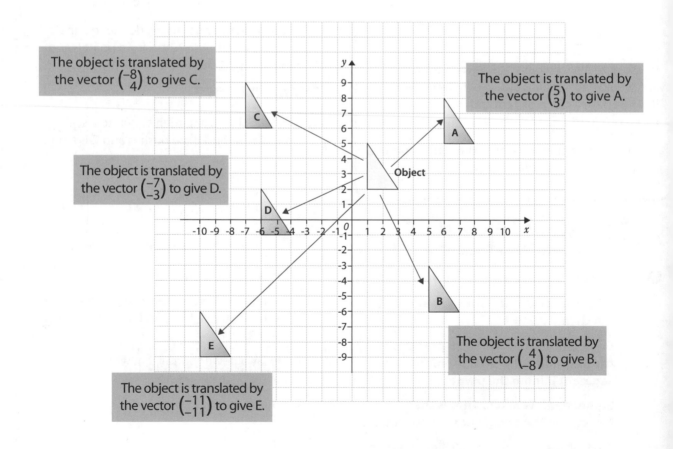

The object is translated by the vector $\begin{pmatrix} -8 \\ 4 \end{pmatrix}$ to give C.

The object is translated by the vector $\begin{pmatrix} 5 \\ 3 \end{pmatrix}$ to give A.

The object is translated by the vector $\begin{pmatrix} -7 \\ -3 \end{pmatrix}$ to give D.

The object is translated by the vector $\begin{pmatrix} 4 \\ -8 \end{pmatrix}$ to give B.

The object is translated by the vector $\begin{pmatrix} -11 \\ -11 \end{pmatrix}$ to give E.

Reflections

Reflections create an image of an object on the other side of the mirror line.

The mirror line is known as an **axis of reflection**.

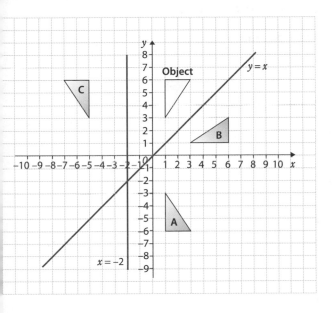

On the example above, the object is reflected in the x-axis (or $y = 0$) to give image A.

The object is reflected in the line $y = x$ to give image B.

The object is reflected in the line $x = -2$ to give image C.

The images and object are congruent.

QUESTIONS

QUICK TEST

1. For the diagram below, describe fully the transformation that maps:

   a. A onto B   **b.** B onto C

   c. A onto D   **d.** A onto E

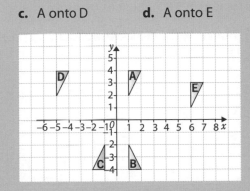

EXAM PRACTICE

1. Triangle A and triangle B have been drawn on the grid below.

   a. Reflect triangle A in the line $x = 5$
   Label this image C.

   b. Translate triangle B by the vector $\begin{pmatrix} 4 \\ 2 \end{pmatrix}$
   Label this image D.

   c. Describe fully the single transformation which will map triangle A onto triangle B.

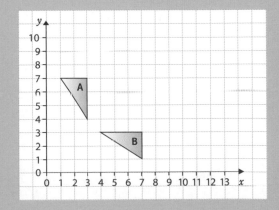

Rotation and Enlargement

Rotations

In a **rotation** the object is turned by a given angle about a fixed point called the **centre of rotation**. The size and shape of the figure are not changed, i.e. the image is **congruent** to the object.

On the example below, object A is rotated by 90° clockwise about (0, 0) to give image B.

Object A is rotated by 180° about (0, 0) to give image C.

Object A is rotated 90° anticlockwise about (−2, 2) to give image D.

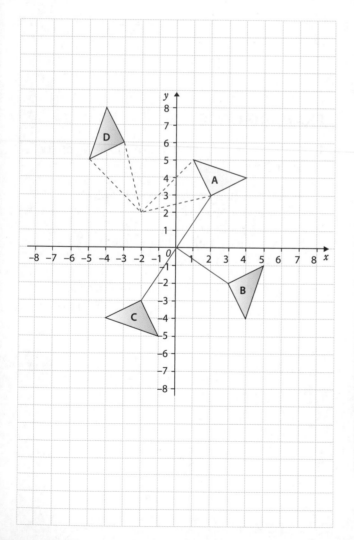

Enlargements

Enlargements change the size but not the shape of the object, i.e. the enlarged shape is **similar** to the object.

The **centre of enlargement** is the point from which the enlargement takes place.

The **scale factor** tells you what all lengths of the original figure have been multiplied by.

Example
Describe fully the transformation that maps ABC onto A'B'C'.

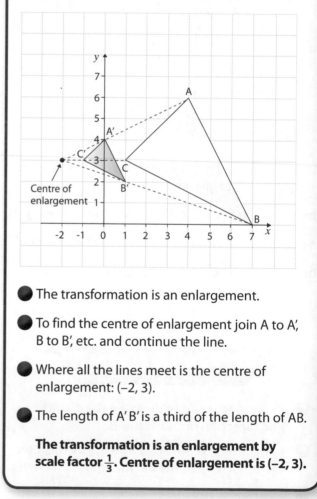

● The transformation is an enlargement.

● To find the centre of enlargement join A to A', B to B', etc. and continue the line.

● Where all the lines meet is the centre of enlargement: (−2, 3).

● The length of A'B' is a third of the length of AB.

The transformation is an enlargement by scale factor $\frac{1}{3}$. Centre of enlargement is (−2, 3).

An enlargement with a scale factor between 0 and 1 makes the shape smaller.

Enlargements with a Negative Scale Factor

For an enlargement with a **negative scale factor**, the image is situated on the opposite side of the centre of enlargement.

In the example below, the triangle A (3, 4), B (9, 4) and C (3, 10) is enlarged with a scale factor of $-\frac{1}{3}$, with the centre of enlargement (0, 1). The enlargement is labelled A'B'C'.

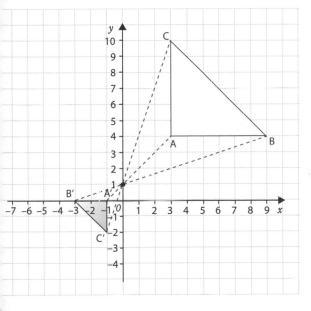

● A'B'C' is on the opposite side of the centre of enlargement from ABC.

● Notice that the length of each side of the triangle A'B'C' is one third the size of the corresponding lengths in the triangle ABC.

SUMMARY

● **After a rotation, the image is congruent to the object.**

● **After an enlargement, the enlarged shape is similar to the object.**

● **An enlargement with a scale factor between 0 and 1 makes the shape smaller.**

● **An enlargement with a negative scale factor means that the image is on the opposite side of the centre of enlargement.**

QUESTIONS

QUICK TEST

1. Complete the diagram below, to show the enlargement of the shape by a scale factor of $\frac{1}{2}$. Centre of enlargement at (0, 0). Call the shape T.

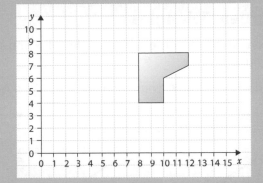

2. Rotate triangle ABC 90° anticlockwise about the point (0, 1). Call the triangle A'B'C'.

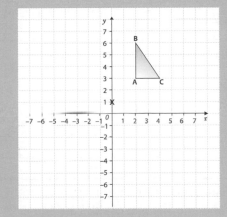

EXAM PRACTICE

1. The quadrilateral is enlarged with a scale factor of $-\frac{1}{2}$ about the origin (0, 0). Draw the enlargement on the diagram. Call the shape A'B'C'D'.

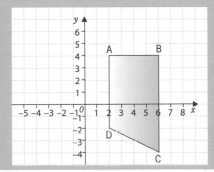

Similarity and Congruency 1

Objects that are exactly the same shape but different sizes are called **similar** shapes. One is an enlargement of the other.

Congruent shapes are identical to each other.

Congruent Triangles

Two triangles are congruent if any of the following sets of conditions is true.

S = side, **A** = angle, **R** = right angle, **H** = hypotenuse.

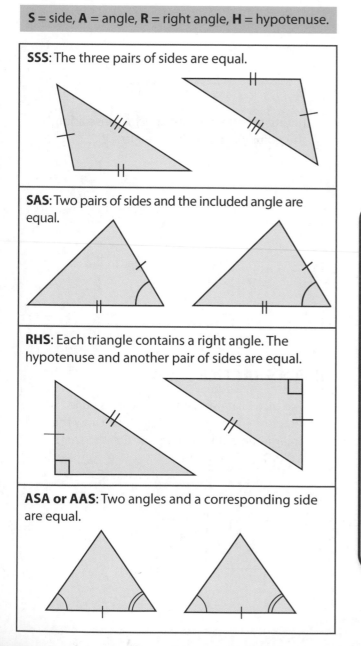

SSS: The three pairs of sides are equal.

SAS: Two pairs of sides and the included angle are equal.

RHS: Each triangle contains a right angle. The hypotenuse and another pair of sides are equal.

ASA or AAS: Two angles and a corresponding side are equal.

Similarity

Questions about finding the missing lengths of similar figures are very common at GCSE.

Remember that corresponding angles are equal. Corresponding lengths are in the same ratio.

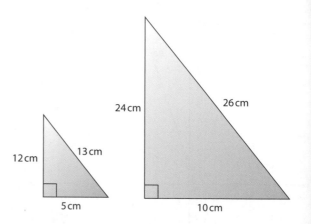

24 cm 26 cm

12 cm 13 cm

5 cm 10 cm

Example

Find the missing length labelled a in the diagram below:

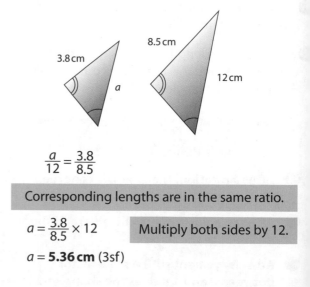

3.8 cm 8.5 cm

a 12 cm

$$\frac{a}{12} = \frac{3.8}{8.5}$$

Corresponding lengths are in the same ratio.

$$a = \frac{3.8}{8.5} \times 12$$ Multiply both sides by 12.

$a = \mathbf{5.36\,cm}$ (3sf)

Examples

1. Find the missing length labelled *a* in the diagram below.

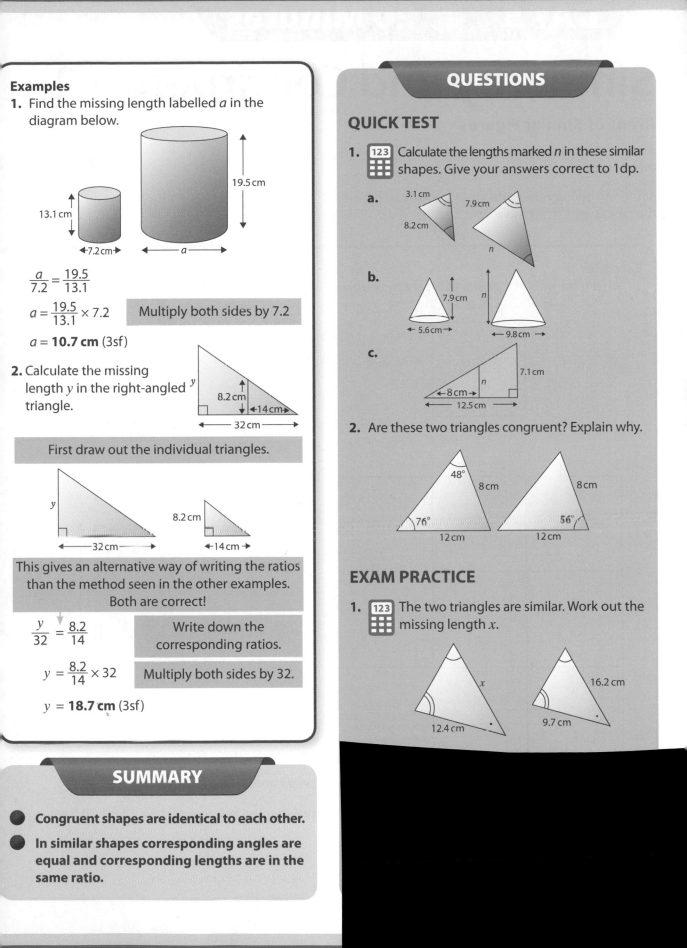

$$\frac{a}{7.2} = \frac{19.5}{13.1}$$

$$a = \frac{19.5}{13.1} \times 7.2 \qquad \boxed{\text{Multiply both sides by 7.2}}$$

$$a = \textbf{10.7 cm } (3sf)$$

2. Calculate the missing length *y* in the right-angled triangle.

> First draw out the individual triangles.

> This gives an alternative way of writing the ratios than the method seen in the other examples. Both are correct!

$$\frac{y}{32} = \frac{8.2}{14} \qquad \boxed{\begin{array}{c}\text{Write down the} \\ \text{corresponding ratios.}\end{array}}$$

$$y = \frac{8.2}{14} \times 32 \qquad \boxed{\text{Multiply both sides by 32.}}$$

$$y = \textbf{18.7 cm } (3sf)$$

QUESTIONS

QUICK TEST

1. 123 Calculate the lengths marked *n* in these similar shapes. Give your answers correct to 1dp.

a. 3.1 cm, 7.9 cm, 8.2 cm, *n*

b. 7.9 cm, *n*, 5.6 cm, 9.8 cm

c. 7.1 cm, *n*, 8 cm, 12.5 cm

2. Are these two triangles congruent? Explain why.

48°, 8 cm, 76°, 12 cm — 8 cm, 56°, 12 cm

EXAM PRACTICE

1. 123 The two triangles are similar. Work out the missing length *x*.

x, 12.4 cm — 16.2 cm, 9.7 cm

SUMMARY

- Congruent shapes are identical to each other.
- In similar shapes corresponding angles are equal and corresponding lengths are in the same ratio.

Similarity and Congruency 2

Areas of Similar Figures

Areas of similar figures are not in the same ratio as their lengths.

If the corresponding lengths are in the ratio $a : b$ then their areas are in the ratio $a^2 : b^2$.

Examples

1. What is the ratio of the areas of the shapes below?

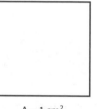

$A = 1 \text{ cm}^2$

2. These two shapes are similar. The area of the larger shape is, Q, 81 cm². The area of the smaller shape, P, is 25 cm². If the height of the larger shape, Q, is 18.9 cm, work out the height of the smaller shape, P.

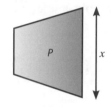

P x

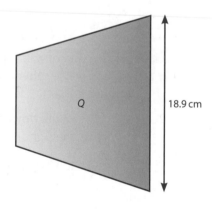

Q 18.9 cm

Area scale factor (k^2) is 25 : 81

Linear scale factor (k) is $\sqrt{25} : \sqrt{81} = 5 : 9$

Height of smaller shape is $\frac{5}{9} \times 18.9 = \mathbf{10.5 \text{ cm}}$

Volumes of Similar Figures

A similar result can be found when looking at the volumes of similar figures.

If the corresponding lengths are in the ratio $a : b$, then their volumes are in the ratio $a^3 : b^3$.

> For a scale factor k:
> The sides are k times bigger
> The areas are k^2 times bigger
> The volumes are k^3 times bigger

Example

These two solids are similar. If the volume of the smaller solid is 9 cm³, calculate the volume of the larger solid.

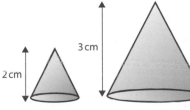

2 cm
3 cm

Linear scale factor is 2 : 3

Volume scale factor is $2^3 : 3^3 = 8 : 27$

Volume of larger solid is $\frac{27}{8} \times 9 = \textbf{30.375 cm}^3$

SUMMARY

● **To find the area of a similar shape:**

– **If the corresponding lengths are in the ratio $a : b$, then their areas are in the ratio $a^2 : b^2$.**

● **To find the volume of a similar shape:**

– **If the corresponding lengths are in the ratio $a : b$, then their volumes are in the ratio $a^3 : b^3$.**

QUESTIONS

QUICK TEST

1. The heights of two similar shapes are 8 cm and 12 cm. If the area of the larger shape is 64 cm², find the area of the smaller shape.

2. [123] Two solids are similar. The heights of the two similar solids are 12 cm and 18 cm. The surface area of the smaller solid is 35 cm². Work out the surface area of the larger solid.

3. [123] Cuboids A and B are similar. The volume of cuboid B is 140 cm³. Calculate the volume of cuboid A.

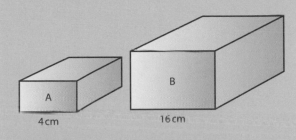

A
4 cm

B
16 cm

Circle Theorems

The Circle Theorems

There are several theorems about circles that you need to know.

1. The perpendicular bisector of any chord passes through the centre of the circle.

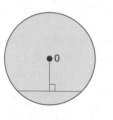

2. The angle in a semicircle is always 90°.

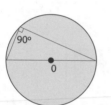

5. The angle at the centre is twice the angle at the circumference subtended by the same arc, e.g. $P\hat{O}Q = 2 \times P\hat{R}Q$.

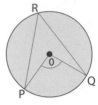

6. Opposite angles of a cyclic quadrilateral add up to 180°.

(A cyclic quadrilateral is a 4-sided shape with each vertex touching the circumference of the circle.)

i.e. $x + y = 180°$

$a + b = 180°$

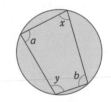

7. The lengths of two tangents from a point are equal i.e. RS = RT.

8. The angle between a tangent and a chord is equal to the angle in the alternate segment (that is the angle which is made at the edge of the circle by two lines drawn from the chord). This is known as the **Alternate Segment Theorem**.

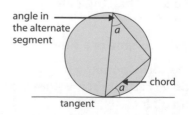

angle in the alternate segment / chord / tangent

Example

Calculate the angles marked *a–d* in the diagram below. Give reasons for your answers.

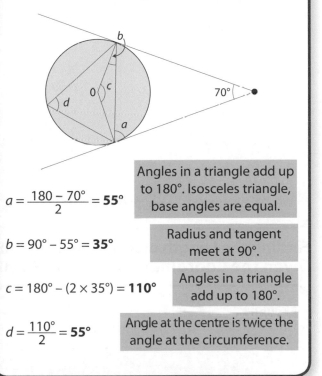

$a = \dfrac{180 - 70°}{2} = \mathbf{55°}$

> Angles in a triangle add up to 180°. Isosceles triangle, base angles are equal.

$b = 90° - 55° = \mathbf{35°}$

> Radius and tangent meet at 90°.

$c = 180° - (2 \times 35°) = \mathbf{110°}$

> Angles in a triangle add up to 180°.

$d = \dfrac{110°}{2} = \mathbf{55°}$

> Angle at the centre is twice the angle at the circumference.

SUMMARY

- The angle between a tangent and radius of a circle is 90°.

- Tangents to a circle from a point outside the circle are equal in length.

- The angles in the same segment subtended by the same arc are equal.

- Opposite angles of a cyclic quadrilateral add up to 180°.

- The angle in a semicircle is a right angle.

- The angle at the centre of a circle is twice the angle at the circumference, both subtended by the same arc.

QUESTIONS

QUICK TEST

1. Some angles are written on cards. Match the missing angles in the diagrams below with the correct card.

   0 represents the centre of the circle.

   53°   50°   62°   109°   126°

   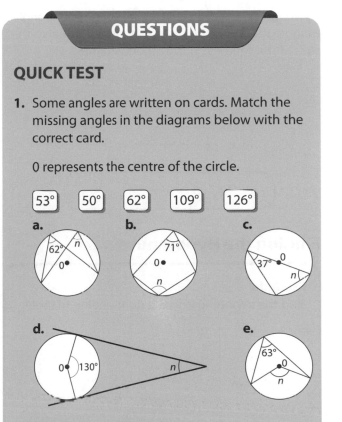

EXAM PRACTICE

1. In the diagram, E, F and G are points on the circle, centre 0.

   Angle FGB = 71°

   AB is a tangent to the circle at point G.

   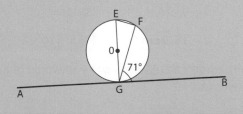

   a. Calculate the size of angle EGF. Give reasons for your answer.

   b. Calculate the size of angle GEF. Give reasons for your answer.

Pythagoras' Theorem

Pythagoras' Theorem states that 'For any right-angled triangle the square on the hypotenuse is equal to the sum of the squares on the other two sides.'

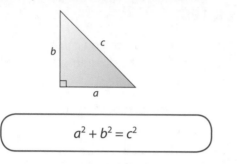

$$a^2 + b^2 = c^2$$

Finding the Hypotenuse

Example

Find the hypotenuse in the right-angled triangle.

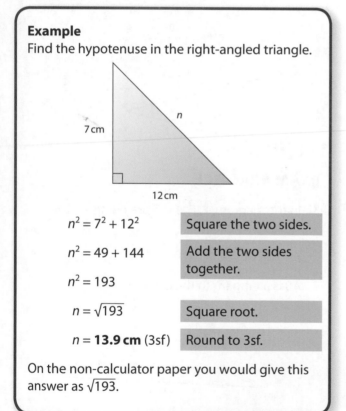

$n^2 = 7^2 + 12^2$	Square the two sides.
$n^2 = 49 + 144$	Add the two sides together.
$n^2 = 193$	
$n = \sqrt{193}$	Square root.
$n = \mathbf{13.9\ cm}$ (3sf)	Round to 3sf.

On the non-calculator paper you would give this answer as $\sqrt{193}$.

Finding a Short Side

Example

Find the length of p.

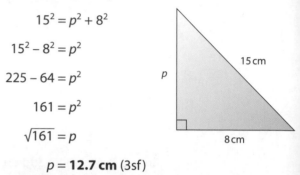

$$15^2 = p^2 + 8^2$$

$$15^2 - 8^2 = p^2$$

$$225 - 64 = p^2$$

$$161 = p^2$$

$$\sqrt{161} = p$$

$$p = \mathbf{12.7\ cm}\ (3sf)$$

When finding a shorter length remember to subtract.

Finding the Length of a Line Segment AB Given the Coordinates of its End Points

Example

Find the length AB in the following diagram.

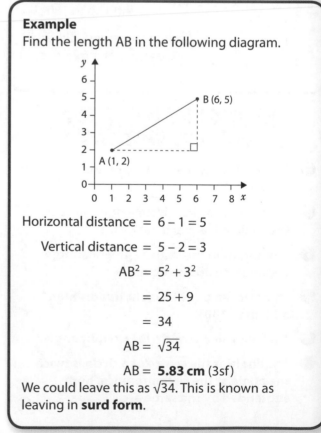

Horizontal distance $= 6 - 1 = 5$

Vertical distance $= 5 - 2 = 3$

$$AB^2 = 5^2 + 3^2$$

$$= 25 + 9$$

$$= 34$$

$$AB = \sqrt{34}$$

$$AB = \mathbf{5.83\ cm}\ (3sf)$$

We could leave this as $\sqrt{34}$. This is known as leaving in **surd form**.

Solving a More Difficult Problem

Calculate the vertical height of this isosceles triangle.

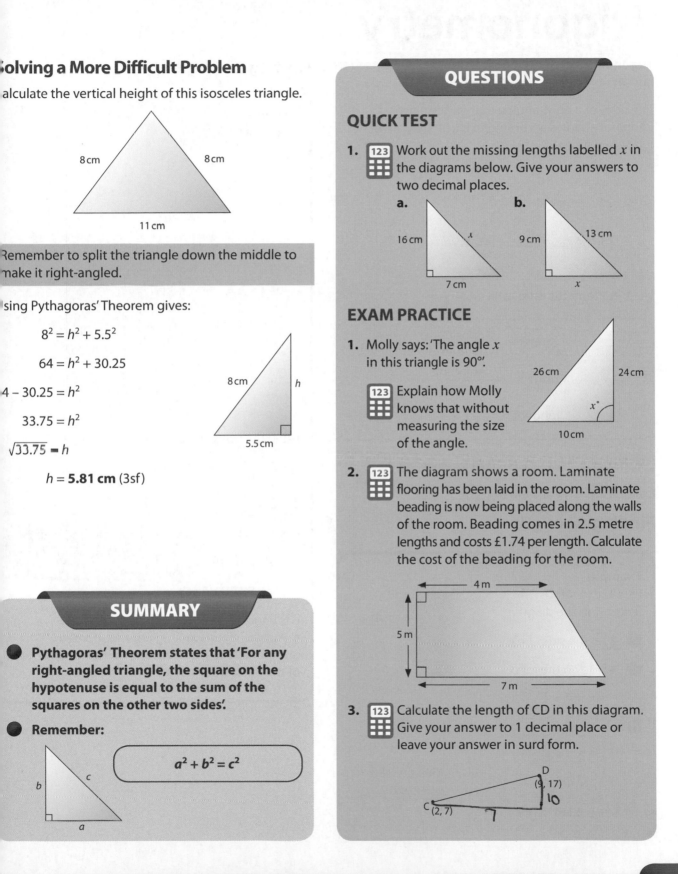

8 cm   8 cm

11 cm

Remember to split the triangle down the middle to make it right-angled.

Using Pythagoras' Theorem gives:

$$8^2 = h^2 + 5.5^2$$

$$64 = h^2 + 30.25$$

$$64 - 30.25 = h^2$$

$$33.75 = h^2$$

$$\sqrt{33.75} = h$$

$$h = \textbf{5.81 cm} \text{ (3sf)}$$

8 cm   h

5.5 cm

SUMMARY

- Pythagoras' Theorem states that 'For any right-angled triangle, the square on the hypotenuse is equal to the sum of the squares on the other two sides'.

- Remember:

$$a^2 + b^2 = c^2$$

b
c
a

QUESTIONS

QUICK TEST

1. [123] Work out the missing lengths labelled x in the diagrams below. Give your answers to two decimal places.

   a.

   16 cm x

   7 cm

   b.

   9 cm   13 cm

   x

EXAM PRACTICE

1. Molly says: 'The angle x in this triangle is 90°'.

   [123] Explain how Molly knows that without measuring the size of the angle.

   26 cm   24 cm

   $x°$

   10 cm

2. [123] The diagram shows a room. Laminate flooring has been laid in the room. Laminate beading is now being placed along the walls of the room. Beading comes in 2.5 metre lengths and costs £1.74 per length. Calculate the cost of the beading for the room.

   4 m

   5 m

   7 m

3. [123] Calculate the length of CD in this diagram. Give your answer to 1 decimal place or leave your answer in surd form.

   D
   (9, 17)
   10
   C (2, 7)
   7

Trigonometry

Trigonometry in right-angled triangles can be used to find an unknown angle or length.

The sides of a right-angled triangle are given temporary names according to where they are in relation to a chosen angle θ.

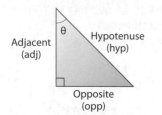

The trigonometric ratios are:

$$\text{Sine } \theta = \frac{\text{Opposite}}{\text{Hypotenuse}}$$

$$\text{Cosine } \theta = \frac{\text{Adjacent}}{\text{Hypotenuse}}$$

$$\text{Tangent } \theta = \frac{\text{Opposite}}{\text{Adjacent}}$$

Use the words **SOH – CAH – TOA** to remember the ratios.

Example: TOA means $\tan \theta = \dfrac{\text{opp}}{\text{adj}}$

Finding a Length

Example
Find the missing length y in the diagram.

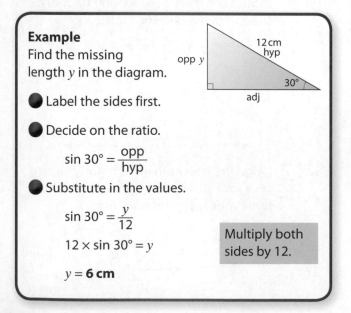

- ● Label the sides first.

- ● Decide on the ratio.

$$\sin 30° = \frac{\text{opp}}{\text{hyp}}$$

- ● Substitute in the values.

$$\sin 30° = \frac{y}{12}$$

$$12 \times \sin 30° = y$$

> Multiply both sides by 12.

$$y = \textbf{6 cm}$$

Finding an Angle

Example
Calculate angle ABC.

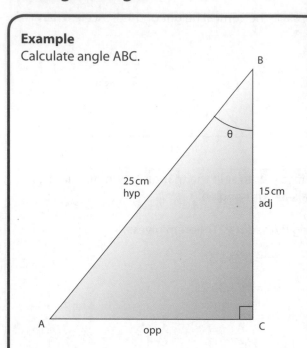

Label the sides and decide on the ratio.

$$\cos \theta = \frac{\text{adj}}{\text{hyp}}$$

$$\cos \theta = \frac{15}{25}$$

$$\cos \theta = 0.6$$

$$\theta = \cos^{-1} 0.6$$

> To find the angle, you usually use the second function on your calculator.

$$= \textbf{53.13°} \text{ (2dp)}$$

It is important you know how to use your calculator when working out trigonometry questions. Also, do not round off before the end of the question.

Angle of Elevation

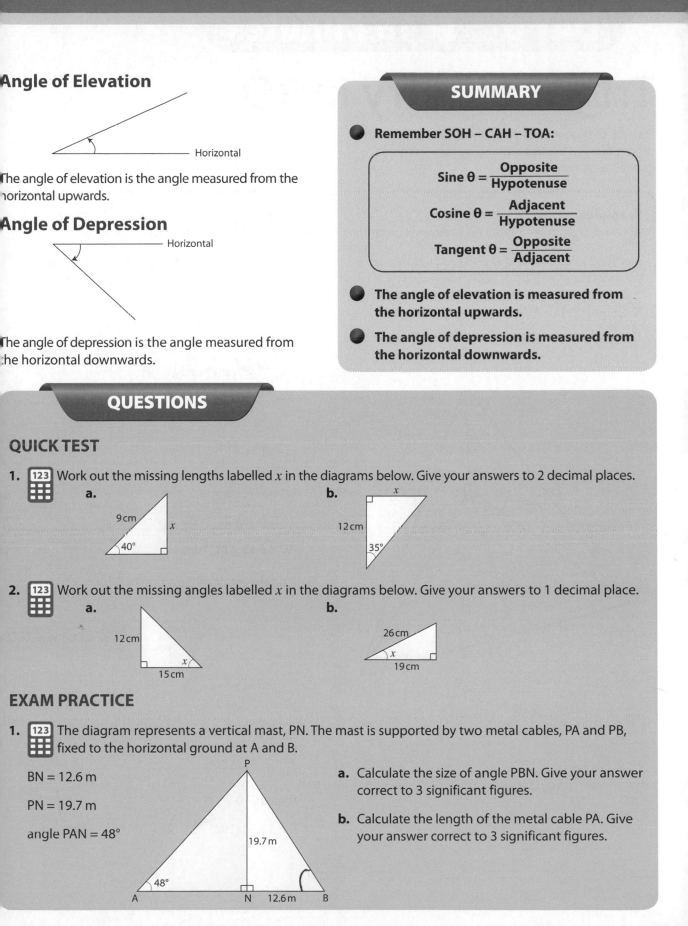

The angle of elevation is the angle measured from the horizontal upwards.

Angle of Depression

The angle of depression is the angle measured from the horizontal downwards.

QUESTIONS

QUICK TEST

1. Work out the missing lengths labelled x in the diagrams below. Give your answers to 2 decimal places.

   a. 9 cm, x, 40°

   b. x, 12 cm, 35°

2. Work out the missing angles labelled x in the diagrams below. Give your answers to 1 decimal place.

   a. 12 cm, 15 cm, x

   b. 26 cm, 19 cm, x

EXAM PRACTICE

1. The diagram represents a vertical mast, PN. The mast is supported by two metal cables, PA and PB, fixed to the horizontal ground at A and B.

   BN = 12.6 m

   PN = 19.7 m

   angle PAN = 48°

   19.7 m, 48°, A, N, 12.6 m, B, P

   a. Calculate the size of angle PBN. Give your answer correct to 3 significant figures.

   b. Calculate the length of the metal cable PA. Give your answer correct to 3 significant figures.

Trigonometry in 3D

The use of trigonometry and Pythagoras' Theorem can be applied when solving problems in three-dimensional figures.

Examples

A, C and D are three points on horizontal ground.

A is 1600 m due west of D and C is 1420 m due south of D.

BD is a vertical tower.

The angle of elevation of B from C is 21°.

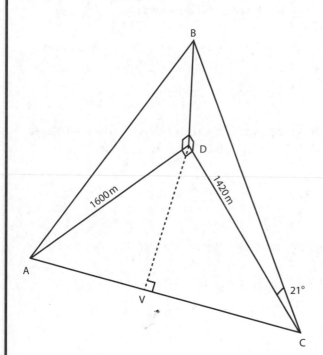

a. Calculate the height of the tower.

> Draw the triangle that you are working with.

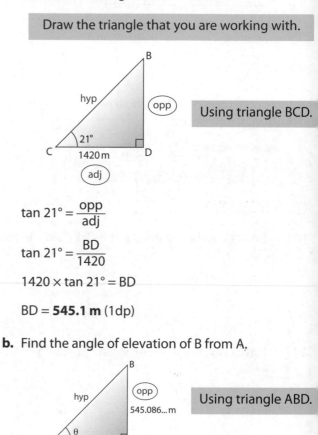

> Using triangle BCD.

$$\tan 21° = \frac{\text{opp}}{\text{adj}}$$

$$\tan 21° = \frac{BD}{1420}$$

$$1420 \times \tan 21° = BD$$

$$BD = \textbf{545.1 m} \ (1dp)$$

b. Find the angle of elevation of B from A.

> Using triangle ABD.

$$\tan \theta = \frac{\text{opp}}{\text{adj}}$$

$$\tan \theta = \frac{545.086\ldots}{1600}$$

$$\theta = \tan^{-1} 0.3406\ldots$$

$$\theta = \textbf{18.8°} \ (1dp)$$

. V is a point on AC which is nearest to D. Calculate the angle of elevation of B from V.

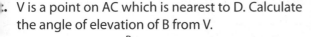

Using triangle BDV.

Since we only have one length in triangle BDV, we need to find some more information.

Using triangle ACD, find angle DCA.

$$\tan \theta = \frac{1600}{1420}$$

$$\theta = \tan^{-1} 1.126 \ldots$$

$$\theta = 48.410 \ldots °$$

Now use triangle DVC to find length DV.

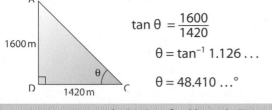

$$\sin 48.41° = \frac{DV}{1420}$$

$$DV = \sin 48.410 \ldots ° \times 1420$$

$$DV = 1062.053 \ldots \text{ m}$$

Now we use triangle BDV to find angle BVD.

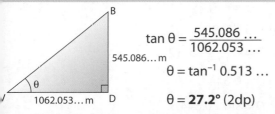

$$\tan \theta = \frac{545.086 \ldots}{1062.053 \ldots}$$

$$\theta = \tan^{-1} 0.513 \ldots$$

$$\theta = \mathbf{27.2°} \text{ (2dp)}$$

This is an example of a multistepped question, in that you have to do several parts to get to the answer. They are usually worth lots of marks.

● When solving problems in 3D, it is useful to draw the right-angled triangle that is needed and apply Pythagoras' Theorem and trigonometry to find missing lengths and angles.

QUESTIONS

QUICK TEST

1. The diagram shows a cuboid.

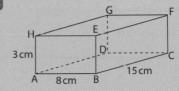

a. Decide whether the following statements are true or false.

i. The length HB is 8.54 cm (2dp).

ii. The length HC is 18 cm.

iii. Angle ACD is 47°.

iv. Angle HBA is 20.6° (1dp).

b. Calculate the size of angle HCA to the nearest degree.

EXAM PRACTICE

1. The diagram shows a square-based pyramid. The point E lies directly above M. M is the centre of the base.

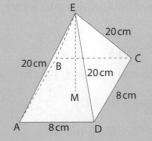

a. Calculate the height of the vertex E above the base. Give your answer to 2 decimal places.

b. Calculate the angle between ED and the base ABCD. Give your answer to 3 significant figures.

Sine and Cosine Rules

The sine and cosine rules allow you to solve problems in triangles that do not contain a right angle.

The Sine Rule

The standard way to write down the sine and cosine rules is to use the following notation for sides and angles in a general triangle.

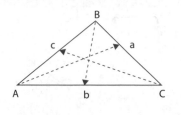

When finding a **length**:

$$\frac{a}{\sin A} = \frac{b}{\sin B} = \frac{c}{\sin C}$$

When finding an **angle**:

$$\frac{\sin A}{a} = \frac{\sin B}{b} = \frac{\sin C}{c}$$

Examples

1. Calculate the length of RS.

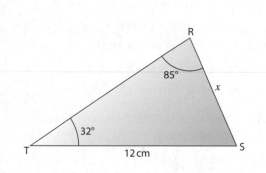

● Call RS, x, the length to be found.

● Since we have two angles and a side, we use the sine rule:
$$\frac{x}{\sin 32°} = \frac{12}{\sin 85°}$$

● Rearrange to make x the subject:
$$x = \frac{12}{\sin 85°} \times \sin 32°$$
Length of RS = **6.38 cm** (3sf)

2. Calculate the size of angle CDE.

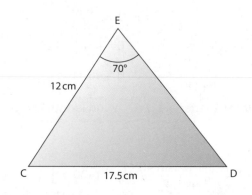

● Since we have two sides and an angle not enclosed by them, we use the sine rule:
$$\frac{\sin D}{12} = \frac{\sin 70°}{17.5}$$

● Rearrange to give:
$$\sin D = \frac{\sin 70°}{17.5} \times 12$$
$$\sin D = 0.644 \dots$$
$$D = \sin^{-1} 0.644\dots$$
$$D = \textbf{40.1°} \text{ (3sf)}$$

The Cosine Rule

When finding a **length**:

$$a^2 = b^2 + c^2 - (2bc \cos A)$$

When finding an **angle**:

$$\cos A = \frac{b^2 + c^2 - a^2}{2bc}$$

Example

1. Calculate the length of JK.

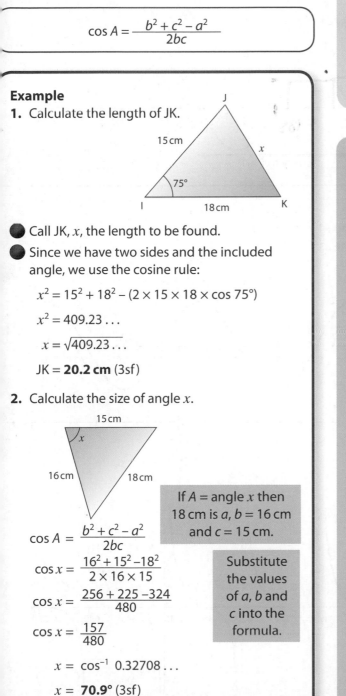

15 cm
x
75°
I
18 cm
K

- Call JK, x, the length to be found.
- Since we have two sides and the included angle, we use the cosine rule:

$$x^2 = 15^2 + 18^2 - (2 \times 15 \times 18 \times \cos 75°)$$

$$x^2 = 409.23 \ldots$$

$$x = \sqrt{409.23 \ldots}$$

$$JK = \textbf{20.2 cm} \text{ (3sf)}$$

2. Calculate the size of angle x.

15 cm
x
16 cm
18 cm

If A = angle x then 18 cm is a, b = 16 cm and c = 15 cm.

$$\cos A = \frac{b^2 + c^2 - a^2}{2bc}$$

$$\cos x = \frac{16^2 + 15^2 - 18^2}{2 \times 16 \times 15}$$

Substitute the values of a, b and c into the formula.

$$\cos x = \frac{256 + 225 - 324}{480}$$

$$\cos x = \frac{157}{480}$$

$$x = \cos^{-1} 0.32708 \ldots$$

$$x = \textbf{70.9°} \text{ (3sf)}$$

SUMMARY

- Sine rule:

  Finding a length: $\dfrac{a}{\sin A} = \dfrac{b}{\sin B} = \dfrac{c}{\sin C}$

  Finding an angle: $\dfrac{\sin A}{a} = \dfrac{\sin B}{b} = \dfrac{\sin C}{c}$

- Cosine rule:

  Finding a length: $a^2 = b^2 + c^2 - (2bc \cos A)$

  Finding an angle: $\cos A = \dfrac{b^2 + c^2 - a^2}{2bc}$

QUESTIONS

QUICK TEST

1. `123` For the questions below, decide whether the missing length x is correct.

   a. $x = 11.49$ cm

   x
   37°
   62°
   7 cm

   b. $x = 20.93$ cm

   x
   12 cm
   107°
   14 cm

2. `123` Calculate the missing angles in the diagrams below to the nearest degree:

   a.

   13 cm
   21 cm
   θ
   12 cm

   b.

   7.4 cm
   10.1 cm
   θ
   36°

EXAM PRACTICE

1. The diagram shows a vertical pole AB on horizontal ground BCD. BCD is a straight line. D is 16 m from C.

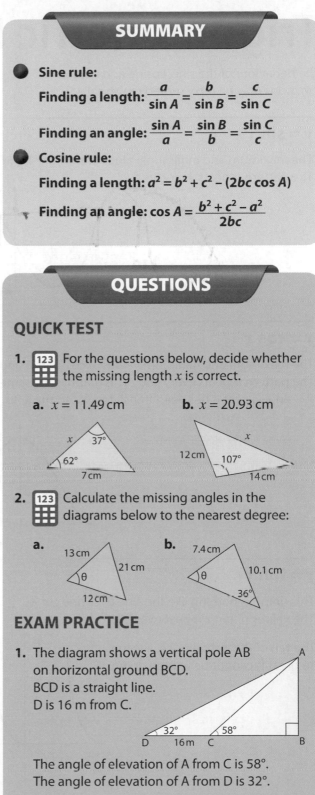

   A
   32°   58°
   D   16 m   C   B

   The angle of elevation of A from C is 58°.
   The angle of elevation of A from D is 32°.

   Calculate the height of the pole.
   Give your answer correct to 3 significant figures.

Trigonometric Functions

The behaviour of the sine, cosine and tangent functions may be represented graphically.

$y = \sin x$

The maximum and minimum values of $\sin x$ are 1 and –1.
The pattern repeats every 360°.

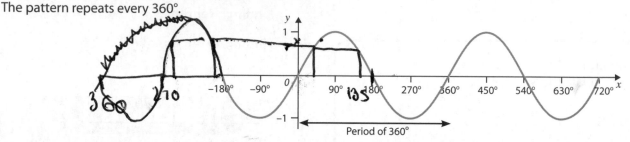

Period of 360°

$y = \cos x$

The maximum and minimum values of $\cos x$ are 1 and –1.
The pattern repeats every 360°. This graph is the same as $y = \sin x$ except it has been translated 90° to the left.

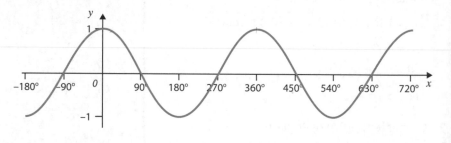

$y = \tan x$

This graph is nothing like the sine or cosine curve.
The values of $\tan x$ repeat every 180°.

The tan of 90° is infinity, i.e. a value so great it cannot be written down.

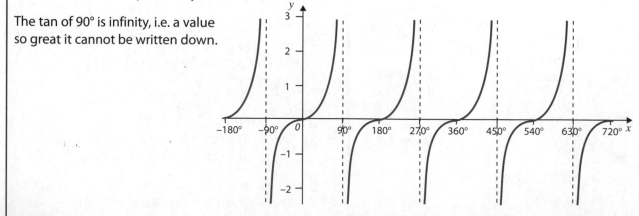

The trigonometric graphs can be used to solve inverse problems.

Example

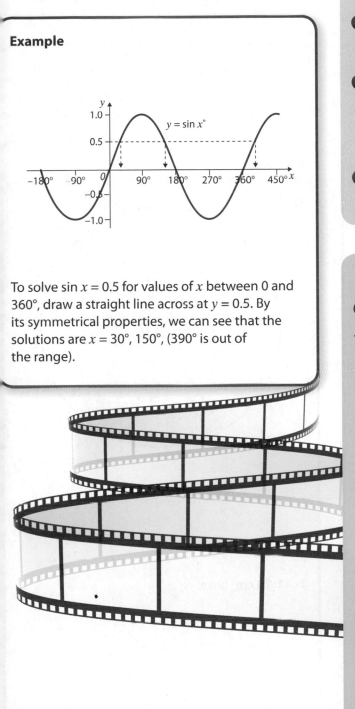

To solve $\sin x = 0.5$ for values of x between 0 and 360°, draw a straight line across at $y = 0.5$. By its symmetrical properties, we can see that the solutions are $x = 30°, 150°$, (390° is out of the range).

SUMMARY

- For the sine graph, the maximum and minimum values of $\sin x$ are 1 and –1. The curve repeats every 360°.

- For the cosine graph, the maximum and minimum values of cosine x are 1 and –1. The curve repeats every 360°. It is the graph of $y = \sin x$ except it has been translated 90° to the left.

- In the graph of $y = \tan x$, the graph repeats every 180°. The value of $\tan 90°$ is infinity.

QUESTIONS

QUICK TEST

1. On the axes below, draw the graph of $y = \sin 2x$.

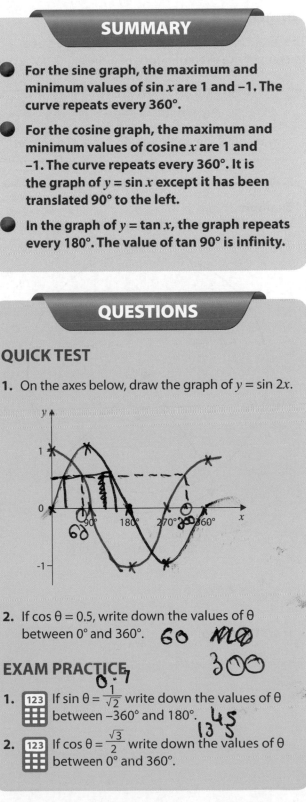

2. If $\cos \theta = 0.5$, write down the values of θ between 0° and 360°. 60 NLØ 3ØØ

EXAM PRACTICE

0.7

1. 123 If $\sin \theta = \frac{1}{\sqrt{2}}$ write down the values of θ between –360° and 180°. 45 135

2. 123 If $\cos \theta = \frac{\sqrt{3}}{2}$ write down the values of θ between 0° and 360°.

Arc, Sector and Segment

Length of a Circular Arc

The length of a circular arc can be expressed as a fraction of the circumference of a circle.

$$\text{Arc length} = \frac{\theta}{360°} \times 2\pi r$$

where θ is the angle subtended at the centre.

Example

Work out the length of the minor arc AB.

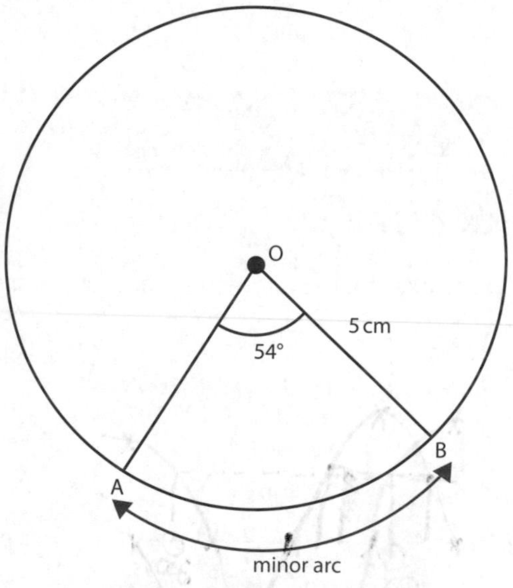

O is the centre of the circle.

$$= \frac{54°}{360°} \times 2 \times \pi \times 5$$

$$= \textbf{4.71 cm} \text{ (2dp)}$$

Area of a Sector

The area of a sector can be expressed as a fraction of the area of a circle.

$$\text{Area of a sector} = \frac{\theta}{360°} \times \pi r^2$$

where θ is the angle subtended at the centre.

Example

Work out the area of the minor sector AOB.

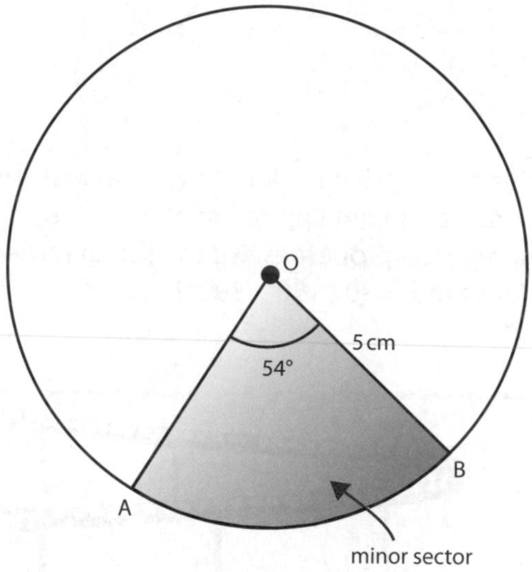

minor sector

$$= \frac{54°}{360°} \times \pi \times 5^2$$

$$= \textbf{11.78 cm}^2 \text{ (2dp)}$$

Area of a General Triangle

If we know the length of two sides of a triangle and the included angle, we can find the area.

$$\text{Area} = \frac{1}{2} \times a \times b \times \sin C$$

two lengths ⟵ included angle

This formula is given on the exam paper.

Example

Work out the area of the triangle below.

$$\text{Area} = \frac{1}{2} \times 10 \times 12 \times \sin 58°$$

$$= \textbf{50.88 cm}^2 \text{ (2dp)}$$

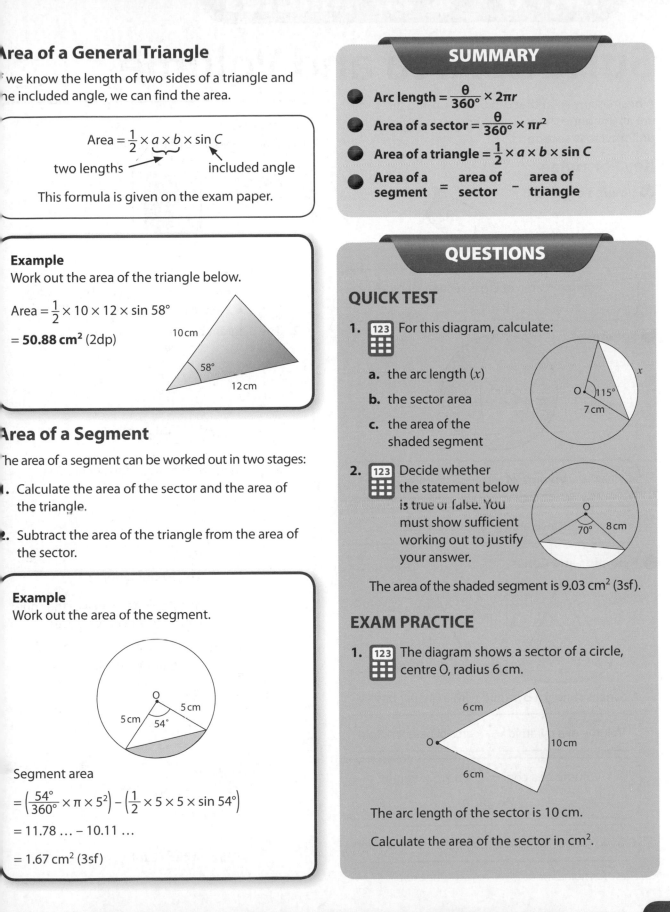

Area of a Segment

The area of a segment can be worked out in two stages:

1. Calculate the area of the sector and the area of the triangle.

2. Subtract the area of the triangle from the area of the sector.

Example

Work out the area of the segment.

Segment area

$$= \left(\frac{54°}{360°} \times \pi \times 5^2\right) - \left(\frac{1}{2} \times 5 \times 5 \times \sin 54°\right)$$

$$= 11.78 \ldots - 10.11 \ldots$$

$$= 1.67 \text{ cm}^2 \text{ (3sf)}$$

QUESTIONS

QUICK TEST

1. [123] For this diagram, calculate:

   a. the arc length (x)

   b. the sector area

   c. the area of the shaded segment

2. [123] Decide whether the statement below is true or false. You must show sufficient working out to justify your answer.

   The area of the shaded segment is 9.03 cm² (3sf).

EXAM PRACTICE

1. [123] The diagram shows a sector of a circle, centre O, radius 6 cm.

   The arc length of the sector is 10 cm.

   Calculate the area of the sector in cm².

Surface Area and Volume

A prism is any solid that can be cut up into slices that are all the same shape. This is known as having a uniform cross-section.

Key Formulae

● **Volume of a Prism**

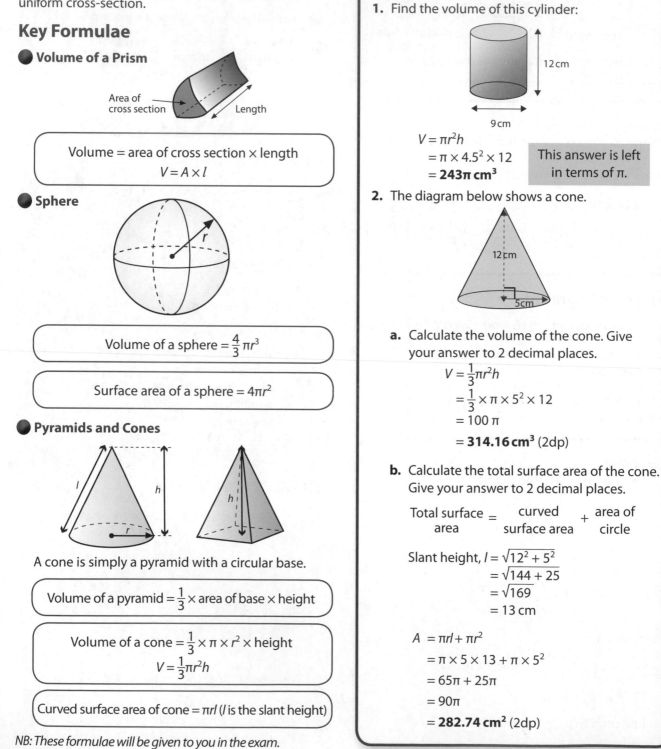

> Area of cross section → ... Length

> Volume = area of cross section × length
> $V = A \times l$

● **Sphere**

> Volume of a sphere = $\frac{4}{3}\pi r^3$

> Surface area of a sphere = $4\pi r^2$

● **Pyramids and Cones**

A cone is simply a pyramid with a circular base.

> Volume of a pyramid = $\frac{1}{3} \times$ area of base × height

> Volume of a cone = $\frac{1}{3} \times \pi \times r^2 \times$ height
> $V = \frac{1}{3}\pi r^2 h$

> Curved surface area of cone = $\pi r l$ (*l* is the slant height)

NB: These formulae will be given to you in the exam.

Examples

1. Find the volume of this cylinder:

12 cm

9 cm

$V = \pi r^2 h$
$= \pi \times 4.5^2 \times 12$
$= \mathbf{243\pi \ cm^3}$

> This answer is left in terms of π.

2. The diagram below shows a cone.

12 cm

5 cm

a. Calculate the volume of the cone. Give your answer to 2 decimal places.

$V = \frac{1}{3}\pi r^2 h$
$= \frac{1}{3} \times \pi \times 5^2 \times 12$
$= 100\pi$
$= \mathbf{314.16 \ cm^3}$ (2dp)

b. Calculate the total surface area of the cone. Give your answer to 2 decimal places.

$$\text{Total surface area} = \text{curved surface area} + \text{area of circle}$$

Slant height, $l = \sqrt{12^2 + 5^2}$
$= \sqrt{144 + 25}$
$= \sqrt{169}$
$= 13$ cm

$A = \pi r l + \pi r^2$
$= \pi \times 5 \times 13 + \pi \times 5^2$
$= 65\pi + 25\pi$
$= 90\pi$
$= \mathbf{282.74 \ cm^2}$ (2dp)

Example

The diagram shows a solid toy. The mass of the toy is 637 grams.

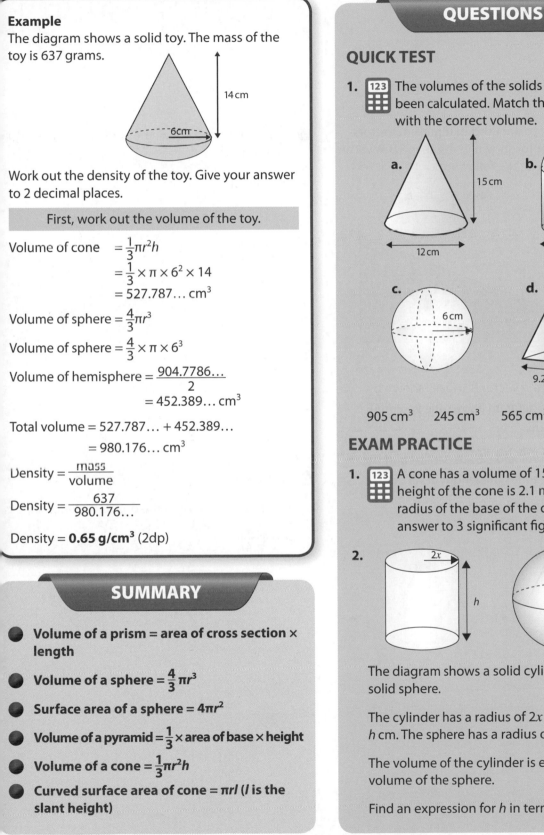

Work out the density of the toy. Give your answer to 2 decimal places.

> First, work out the volume of the toy.

Volume of cone $= \frac{1}{3}\pi r^2 h$

$\qquad = \frac{1}{3} \times \pi \times 6^2 \times 14$

$\qquad = 527.787\ldots \text{ cm}^3$

Volume of sphere $= \frac{4}{3}\pi r^3$

Volume of sphere $= \frac{4}{3} \times \pi \times 6^3$

Volume of hemisphere $= \dfrac{904.7786\ldots}{2}$

$\qquad = 452.389\ldots \text{ cm}^3$

Total volume $= 527.787\ldots + 452.389\ldots$

$\qquad = 980.176\ldots \text{ cm}^3$

Density $= \dfrac{\text{mass}}{\text{volume}}$

Density $= \dfrac{637}{980.176\ldots}$

Density $= \mathbf{0.65 \text{ g/cm}^3}$ (2dp)

SUMMARY

- Volume of a prism = area of cross section × length

- Volume of a sphere $= \frac{4}{3}\pi r^3$

- Surface area of a sphere $= 4\pi r^2$

- Volume of a pyramid $= \frac{1}{3} \times$ area of base × height

- Volume of a cone $= \frac{1}{3}\pi r^2 h$

- Curved surface area of cone $= \pi r l$ (*l* is the slant height)

QUESTIONS

QUICK TEST

1. **123** The volumes of the solids below have been calculated. Match the correct solid with the correct volume.

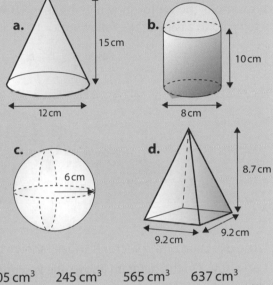

$905 \text{ cm}^3 \qquad 245 \text{ cm}^3 \qquad 565 \text{ cm}^3 \qquad 637 \text{ cm}^3$

EXAM PRACTICE

1. **123** A cone has a volume of 15 m³. The vertical height of the cone is 2.1 m. Calculate the radius of the base of the cone. Give your answer to 3 significant figures.

2.

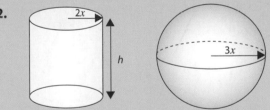

The diagram shows a solid cylinder and a solid sphere.

The cylinder has a radius of $2x$ cm and a height, h cm. The sphere has a radius of $3x$ cm.

The volume of the cylinder is equal to the volume of the sphere.

Find an expression for h in terms of x.

Vectors 1

Vectors

A vector is a quantity that has both distance and direction.

Four types of notation are used to represent vectors. For example, the vector shown in the shape below can be referred to as any of the following:

$$\begin{pmatrix} 5 \\ 2 \end{pmatrix} \qquad \underline{a} \qquad \overrightarrow{AB} \qquad \mathbf{a}$$

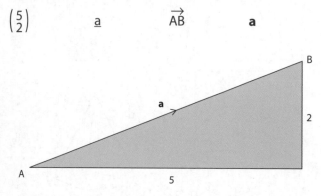

The direction of the vector is usually shown by an arrow. On the diagram above, the vector $\overrightarrow{AB}$ is shown by an arrow.

⬤ If $\overrightarrow{DE} = k\overrightarrow{AB}$, then $\overrightarrow{AB}$ and $\overrightarrow{DE}$ are parallel and the length of $\overrightarrow{DE}$ is k times the length of $\overrightarrow{AB}$.

$$\overrightarrow{AB} = \begin{pmatrix} 5 \\ 2 \end{pmatrix}$$

$$\overrightarrow{DE} = \begin{pmatrix} 10 \\ 4 \end{pmatrix} = 2\begin{pmatrix} 5 \\ 2 \end{pmatrix}$$

$$\overrightarrow{DE} = 2\overrightarrow{AB}$$

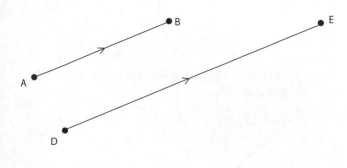

⬤ If two vectors are equal, they are parallel and equal in length.

$\overrightarrow{AB}$ is equal to **a**.

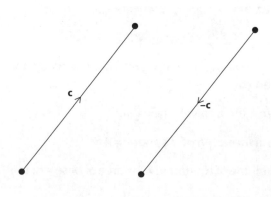

⬤ If the vector $\mathbf{c} = \begin{pmatrix} 3 \\ 4 \end{pmatrix}$

then the vector $-\mathbf{c}$ is in the opposite direction to **c**.

$$-\mathbf{c} = \begin{pmatrix} -3 \\ -4 \end{pmatrix}$$

Magnitude of a Vector

The magnitude of a vector is the length of the directed line segment representing it.

Pythagoras' Theorem can be used to find the magnitude.

In general, the magnitude of a vector $\begin{pmatrix} x \\ y \end{pmatrix}$ is $\sqrt{x^2 + y^2}$.

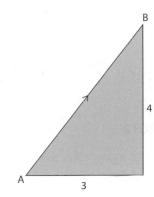

The magnitude of AB:

$$AB = \sqrt{3^2 + 4^2}$$

$$= \sqrt{9 + 16}$$

$$= \sqrt{25}$$

$$AB = 5$$

The magnitude of AB is 5 units.

QUICK TEST

1. Vector $\mathbf{b} = \begin{pmatrix} 4 \\ -2 \end{pmatrix}$. Draw vector **b**.

2. Vector $\mathbf{c} = \begin{pmatrix} -4 \\ 6 \end{pmatrix}$ and vector $\mathbf{d} = \begin{pmatrix} -8 \\ 12 \end{pmatrix}$.
   Explain whether the vectors are parallel.

3. Work out the magnitude of the following vectors. Leave your answer in surd form, where necessary.

   a. $\begin{pmatrix} 4 \\ 2 \end{pmatrix}$

   b. $\begin{pmatrix} -3 \\ 4 \end{pmatrix}$

   c. $\begin{pmatrix} -2 \\ 0 \end{pmatrix}$

   d. $\begin{pmatrix} 6 \\ -1 \end{pmatrix}$

EXAM PRACTICE

1. The diagram is a sketch.
   A is the point (3, 2)
   B is the point (7, 7)

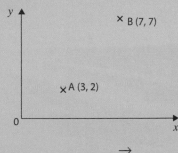

a. Write down the vector $\overrightarrow{AB}$.
   Write your answer as a column vector $\begin{pmatrix} x \\ y \end{pmatrix}$.

b. Write down the coordinates of point C such that $\overrightarrow{BC} = \begin{pmatrix} -2 \\ -3 \end{pmatrix}$.

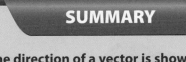

SUMMARY

● **The direction of a vector is shown with an arrow.**

● **The magnitude of a vector is the length of the directed line segment representing it.**

Vectors 2

Addition and Subtraction of Vectors

The **resultant** of two vectors is found by adding them.

Vectors must always be added end to end so that the arrows follow on from each other.

A resultant is usually labelled with a double arrow.

Addition

The triangle below shows the triangle law of addition.

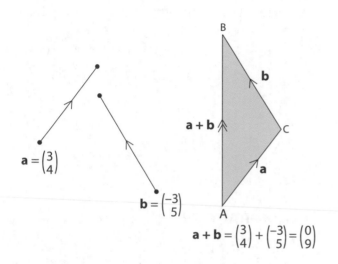

$$\mathbf{a} = \begin{pmatrix} 3 \\ 4 \end{pmatrix}$$

$$\mathbf{b} = \begin{pmatrix} -3 \\ 5 \end{pmatrix}$$

$$\mathbf{a} + \mathbf{b} = \begin{pmatrix} 3 \\ 4 \end{pmatrix} + \begin{pmatrix} -3 \\ 5 \end{pmatrix} = \begin{pmatrix} 0 \\ 9 \end{pmatrix}$$

To take the route directly from A to B is equivalent to travelling via C. Hence we can represent $\overrightarrow{AB}$ as **a** + **b**.

Subtraction

Vectors can also be subtracted.

For example, in the diagram below:

a − **b** can be interpreted as **a** + (−**b**).

$$\mathbf{a} + (-\mathbf{b}) = \begin{pmatrix} 3 \\ 4 \end{pmatrix} + \begin{pmatrix} 3 \\ -5 \end{pmatrix}$$

$$\therefore \ \mathbf{a} - \mathbf{b} = \begin{pmatrix} 6 \\ -1 \end{pmatrix}$$

Position Vectors

The position vector of a point B is the vector $\overrightarrow{OB}$, where O is the origin.

In the diagram, the position vectors of B and C are **b** and **c** respectively. Using this notation:

$$\overrightarrow{BC} = -\mathbf{b} + \mathbf{c} = \mathbf{c} - \mathbf{b}$$

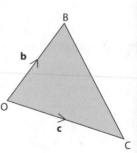

Example

OAB is a triangle. Given that $\overrightarrow{OA} = \mathbf{a}$, $\overrightarrow{OB} = \mathbf{b}$ and that N splits $\overrightarrow{AB}$ in the ratio 1 : 2, find in terms of $\mathbf{a}$ and $\mathbf{b}$ the vectors:

i. $\overrightarrow{AB}$ **ii.** $\overrightarrow{ON}$

i. $\overrightarrow{AB} = \overrightarrow{AO} + \overrightarrow{OB}$ Go from A to B via O.

$\qquad = -\mathbf{a} + \mathbf{b}$

ii. $\overrightarrow{ON} = \overrightarrow{OA} + \overrightarrow{AN}$ $\overrightarrow{AN} = \frac{1}{3}\overrightarrow{AB}$

$\qquad = \mathbf{a} + \frac{1}{3}(-\mathbf{a} + \mathbf{b})$

$\qquad = \mathbf{a} - \frac{1}{3}\mathbf{a} + \frac{1}{3}\mathbf{b}$

$\qquad = \frac{2}{3}\mathbf{a} + \frac{1}{3}\mathbf{b}$

$\qquad = \frac{1}{3}(2\mathbf{a} + \mathbf{b})$

SUMMARY

● The resultant of two vectors is found by vector addition or subtraction.

● Vectors must always be combined end to end.

QUESTIONS

QUICK TEST

1. OABC is a parallelogram.
AB is parallel to OC.
OA is parallel to CB.

$\overrightarrow{OA} = \mathbf{a}$

$\overrightarrow{OC} = \mathbf{c}$

  a. Express in terms of $\mathbf{a}$ and $\mathbf{c}$:

    i. $\overrightarrow{AC}$

    ii. $\overrightarrow{BO}$

  b. N is the midpoint of $\overrightarrow{AC}$:

    Express $\overrightarrow{ON}$ in terms of $\mathbf{a}$ and $\mathbf{c}$.

EXAM PRACTICE

1. ABCDEF is a regular hexagon.

Given that $\overrightarrow{OB} = \mathbf{a}$ and $\overrightarrow{OC} = \mathbf{b}$:

  a. find in terms of $\mathbf{a}$ and $\mathbf{b}$ the vectors:

    i. $\overrightarrow{BC}$

    ii. $\overrightarrow{AD}$

  b. Write down the vector $\overrightarrow{FE}$.

  c. What geometrical fact is exhibited by the vectors $\overrightarrow{FE}$ and $\overrightarrow{AD}$?

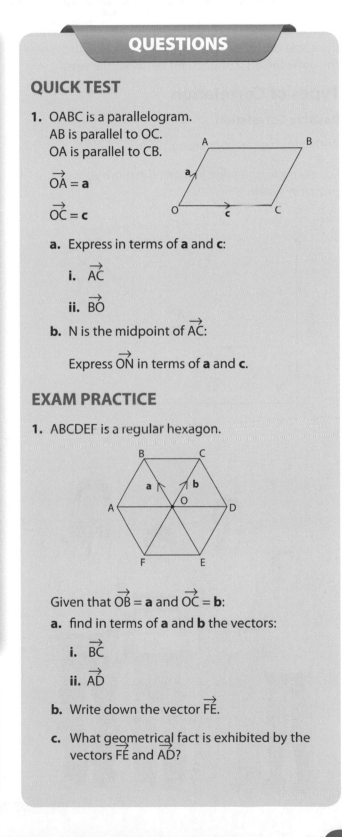

Scatter Diagrams and Correlation

Scatter diagrams are used to show two sets of data at the same time. They are important because they show the **correlation** (connection) between the sets.

Types of Correlation

Positive Correlation

Both variables are increasing.

For example, the taller you are, the more you probably weigh.

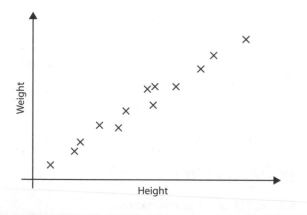

Negative Correlation

As one variable increases, the other decreases.

For example, as the temperature increases, the sale of woollen hats probably decreases.

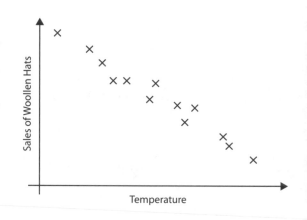

Zero Correlation

Little or no linear relationship between the variables.

For example, there is no connection between your height and your mathematical ability.

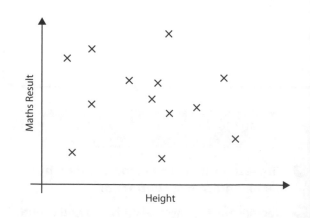

Line of Best Fit

The **line of best** fit goes as close as possible to all the points.

There is roughly an equal number of points above the line and below it.

The scatter diagram below shows the Science and Maths percentages scored by some students.

- The line of best fit goes in the direction of the data.
- The line of best fit can be used to estimate results.
- We can estimate that a student with a Science percentage of 30 would get a Maths percentage of about 17.
- We can estimate that a student with a Maths percentage of 50 would get a Science percentage of about 54.

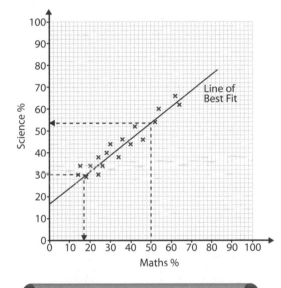

SUMMARY

- **Positive correlation: both variables are increasing.**
- **Negative correlation: as one variable increases, the other decreases.**
- **Zero correlation: little or no linear correlation between the variables.**
- **A line of best fit should be as close as possible to all the points. It is in the direction of the data.**

QUESTIONS

QUICK TEST

Decide whether these statements are **true** or **false**.

1. There is a positive correlation between the weight of a book and the number of pages.

2. There is no correlation between the height you climb up a mountain and the temperature.

3. There is a negative correlation between the age of a used car and its value.

4. There is a positive correlation between the height of some students and the size of their feet.

5. There is no correlation between the weight of some students and their History GCSE results.

EXAM PRACTICE

1. The scatter diagram shows the age of some cars and their values.

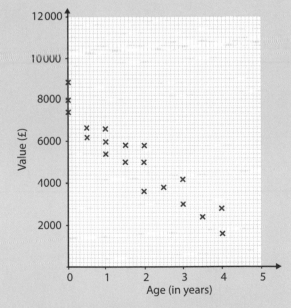

a. Draw a line of best fit on the diagram.

b. Use your line of best fit to estimate the age of a car when its value is £5000.

c. Use your line of best fit to estimate the value of a $3\frac{1}{2}$ year-old car.

Averages

Averages of Continuous Data

When data is grouped into **class intervals**, the exact data is not known.

We estimate the **mean** by using the midpoints of the class intervals.

> Add in 2 extra columns – one for the midpoint and one for fx.

Weight (W kg)	Frequency (f)	Midpoint (x)	fx
$30 \leqslant W < 35$	6	32.5	195
$35 \leqslant W < 40$	14	37.5	525
$40 \leqslant W < 45$	22	42.5	935
$45 \leqslant W < 50$	18	47.5	855
	60		**2510**

↑ Σf ↑ Σfx

For continuous data:

$$\bar{x} = \frac{\Sigma fx}{\Sigma f}$$

Σ means the sum of
f represents the frequency
$\bar{x}$ represents the mean
x represents the midpoint of the class interval

$$\bar{x} = \frac{\Sigma fx}{\Sigma f}$$

$$\bar{x} = \frac{2510}{60}$$

$$\bar{x} = 41.8\dot{3}$$

Modal class is $40 \leqslant W < 45$

This class interval has the highest frequency.

To find the class interval containing the **median**, first find the position of the median:

$$= \frac{\Sigma f + 1}{2}$$

$$= \frac{60 + 1}{2} = 30.5$$

The median lies halfway between the 30th and 31st values. The 30th and 31st values are in the class interval $40 \leqslant W < 45$.

Hence, the class interval in which the median lies is $40 \leqslant W < 45$.

Stem and Leaf Diagrams

Stem and leaf diagrams are useful for recording and displaying information. They can also be used to find the mode, median and range of a set of data.

These are the marks gained by some students in a Maths test:

52	45	63	67
75	57	68	67
60	59	67	

In an ordered stem and leaf diagram, it would look like this

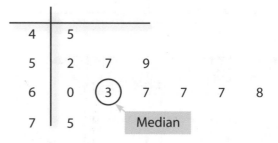

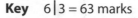

Key $6 \mid 3 = 63$ marks

The median is the sixth value = 63 marks

The mode is 67 marks

The range (highest value – lowest value) is $75 - 45 = 30$ marks

f the same students sat a second Maths test, their
results could be put into a **back-to-back stem and
leaf diagram**. These are very useful when comparing
two sets of data.

Test 2				Test 1					
9	6	2	4	5					
7	5	1	1	5	2	7	9		
	3	1	0	6	0	3	7	7	7	8
			7	5					

Key Test 1 $5 \mid 2 = 52$ marks
    Test 2 $1 \mid 5 = 51$ marks

Comparing the data in the back-to-back stem and leaf
diagram, we can say that:

In test 2, the median score of 55 marks is lower than
the median score in test 1 of 63 marks. The range of
the scores in test 2 is 21 marks, which is lower than
the range of the scores in test 1 of 30 marks. So on
average, the students did better in test 1 but their
scores were more variable than in test 2.

SUMMARY

● **For continuous data:**

  – **Estimate the mean using:**

  $$\bar{x} = \frac{\Sigma fx}{\Sigma f}$$

  **Σ means the sum of
  f represents the frequency
  $\bar{x}$ represents the mean
  x represents the midpoint of
  the class interval**

  – **Modal class is the class interval with the
    highest frequency.**

● **Use stem and leaf diagrams to record and
  display information. Do not forget to order
  the leaves and write a key.**

● **Back-to-back stem and leaf diagrams can
  be used to compare two sets of data.**

QUESTIONS

QUICK TEST

1. 🔢 The heights, h cm, of some students are
   shown in the table.

Height (h cm)	Frequency	Midpoint	fx
$140 \leqslant h < 145$	4		
$145 \leqslant h < 150$	9		
$150 \leqslant h < 155$	15		
$155 \leqslant h < 160$	6		

Calculate an estimate for the mean of this data.

2. **a.** Draw an accurate stem and leaf diagram of
   this data.

27	28	36	42	50	18
25	31	39	25	49	31
33	27	37	25	47	40
7	31	26	36	9	42

   b. What is the median of this data?

EXAM PRACTICE

1. The table shows information about the number
   of hours that 50 children watched television for
   last week.

   Work out an estimate for the mean number
   of hours the children watched television.

Number of Hours (h)	Frequency
$0 \leqslant h < 2$	3
$2 \leqslant h < 4$	6
$4 \leqslant h < 6$	22
$6 \leqslant h < 8$	13
$8 \leqslant h < 10$	6

Cumulative Frequency Graphs

With a **cumulative frequency graph**, it is possible to estimate the median of grouped data and the interquartile range.

Example

The cumulative frequency table opposite shows the marks of 94 students in a Maths exam.

a. Complete the cumulative frequency table for this data.

b. Draw a cumulative frequency graph for this data.

To do this we must plot the upper boundary of each class interval on the x-axis and the cumulative frequency on the y-axis.
Plot (20, 2) (30, 8) (40, 18) …
Join the points with a smooth curve.
Since no students had less than zero marks, the graph starts at (0, 0).

c. Use the cumulative frequency graph to find the median and interquartile range of the data.

> Cumulative frequency is a running total of all the frequencies.

Mark (m)	Frequency	Mark	Cumulative Frequency
$0 < m \leqslant 20$	2	$\leqslant 20$	2 (2)
$20 < m \leqslant 30$	6	$\leqslant 30$	8 (2 + 6)
$30 < m \leqslant 40$	10	$\leqslant 40$	18 (8 + 10)
$40 < m \leqslant 50$	17	$\leqslant 50$	35 (18 + 17)
$50 < m \leqslant 60$	24	$\leqslant 60$	59 (35 + 24)
$60 < m \leqslant 70$	17	$\leqslant 70$	76 (59 + 17)
$70 < m \leqslant 80$	11	$\leqslant 80$	87 (76 + 11)
$80 < m \leqslant 90$	4	$\leqslant 90$	91 (87 + 4)
$90 < m \leqslant 100$	3	$\leqslant 100$	94 (91 + 3)

> This means that 94 students had a score of 100 or less.

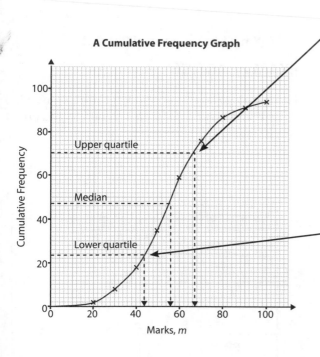

A Cumulative Frequency Graph

The **upper quartile** is three quarters of the way into the distribution:
$$\frac{3}{4} \times 94 = 70.5$$
Read across from 70.5 and down to the horizontal axis.
Upper quartile ≈ 67 marks

The **median** splits the data into two halves – the lower 50% and the upper 50%.
$$\text{Median} = \frac{1}{2} \times \text{cumulative frequency}$$
$$= \frac{1}{2} \times 94 = 47$$
Read across from 47. Median ≈ 56 marks

The **lower quartile** is the value one quarter of the way into the distribution:
$$\frac{1}{4} \times 94 = 23.5$$
Read across from 23.5. Lower quartile ≈ 44 marks

Interquartile range
= upper quartile – lower quartile
= 67 – 44 = 23 marks

A large interquartile range indicates that the 'middle half' of the data is widely spread about the median.

A small interquartile range indicates that the 'middle half' of the data is concentrated about the median.

Box Plots

- Box plots are sometimes known as box and whisker diagrams.

- Cumulative frequency graphs are not easy to compare; a box plot shows the interquartile range as a box and the highest and lowest values as whiskers. Comparing the spread of data is then easier.

Example
The box plot of the cumulative frequency graph opposite would look like this:

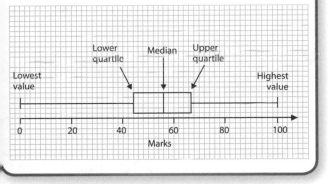

QUESTIONS

QUICK TEST

1. The box plot below shows the times in minutes to finish an assault course.

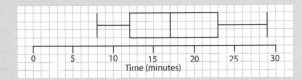

Use the box plot to complete the table below.

	Time in Minutes
Median time	
Lower quartile	
Interquartile range	
Longest time	

EXAM PRACTICE

1. Students in 9A and 9B took the same test. Their results were used to draw the following box plots.

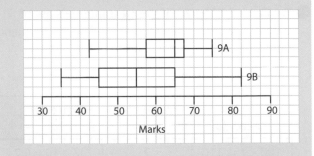

a. In which class was the student who scored the highest mark?

b. In which class did the students perform better in the test? You must give a reason for your answer.

Histograms

In a **histogram** the area of a bar is proportional to the frequency.

- If the class intervals have **equal widths**, frequency can be used for the height of the bar.

- If the class intervals have **unequal widths**, the height of the bar is adjusted by using **frequency density** and the area of the bar is equal to the frequency.

$$\text{Frequency density} = \frac{\text{frequency}}{\text{class width}}$$

The **modal class** is the class interval with the highest bar.

Drawing Histograms

The table below shows the time in seconds it takes people to swim 100 metres.

Time, t (seconds)	Frequency
$100 < t \leqslant 110$	2
$110 < t \leqslant 140$	24
$140 < t \leqslant 160$	42
$160 < t \leqslant 200$	50
$200 < t \leqslant 220$	24
$220 < t \leqslant 300$	20

- To draw a histogram if the class intervals are of different widths, you need to calculate the frequency densities. Add an extra column to the table.

The table should now look like this:

Time, t (seconds)	Frequency	Frequency Density
$100 < t \leqslant 110$	2	$2 \div 10 = 0.2$
$110 < t \leqslant 140$	24	$24 \div 30 = 0.8$
$140 < t \leqslant 160$	42	$42 \div 20 = 2.1$
$160 < t \leqslant 200$	50	$50 \div 40 = 1.25$
$200 < t \leqslant 220$	24	$24 \div 20 = 1.2$
$220 < t \leqslant 300$	20	$20 \div 80 = 0.25$

- Draw on graph paper – make sure there are no gaps between bars.

The histogram should look like this:

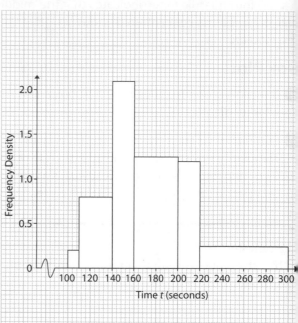

- Sometimes you will be asked to read from a histogram. In this case rearrange the formula for frequency density.

$$\text{Frequency} = \text{frequency density} \times \text{class width}$$

QUICK TEST

1. The table opposite gives some information about the ages of participants in a charity walk.

   On graph paper draw a histogram to represent this information.

Age (x) in years	Frequency
$0 < x \leq 15$	30
$15 < x \leq 25$	46
$25 < x \leq 40$	45
$40 < x \leq 60$	25

EXAM PRACTICE

1. The table and histogram give information about the distance (d km) travelled to work by some employees.

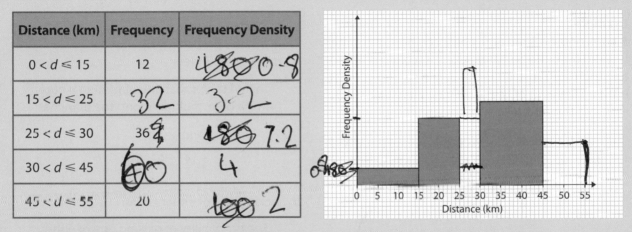

Distance (km)	Frequency	Frequency Density
$0 < d \leq 15$	12	~~48~~ ~~0~~ 0·8
$15 < d \leq 25$	32	3·2
$25 < d \leq 30$	36 ~~8 4~~	~~18~~ 7·2
$30 < d \leq 45$	~~60~~	4
$45 < d \leq 55$	20	~~100~~ 2

   a. Use the information in the histogram to complete the table.

   b. Use the table to complete the histogram.

2. The histogram shows the masses of onions in grams in a sack.

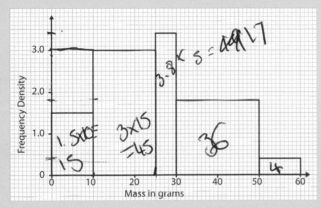

   a. How many onions have a mass of between 10 and 25 grams?

   b. How many onions were there in total?

Probability

Tree diagrams are used to show the possible outcomes of two or more events. There are two rules you need to know first.

● **The OR rule**
If two events are mutually exclusive, the probability of A or B happening is found by adding the probabilities.

$$P(A \text{ or } B) = P(A) + P(B)$$

● **The AND rule**
If two events are independent, the probability of A and B happening together is found by multiplying the separate probabilities.

$$P(A \text{ and } B) = P(A) \times P(B)$$

(These rules also work for more than two events.)

Example
A bag contains 3 red and 4 blue counters. A counter is taken from the bag at random, its colour is noted and then it is replaced in the bag. A second counter is then taken out of the bag. Draw a tree diagram to illustrate this information.

Remember that the probabilities on the branches leaving each point on the tree add up to 1.

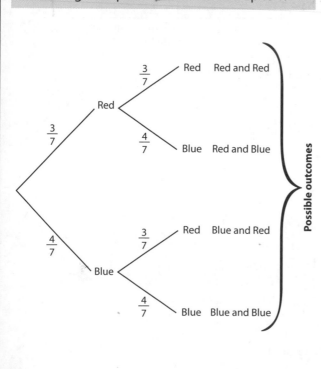

Work out the probability of:
a. picking two blues

To find the probability of picking a blue AND a blue, multiply along the branches.

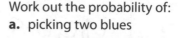
$P(B) \times P(B)$
$$= \frac{4}{7} \times \frac{4}{7} = \frac{16}{49}$$

b. picking one of either colour

The probability of picking one of either colour is the probability of picking a blue and a red OR a red and a blue. You need to use the AND rule and the OR rule.

P(blue and red)
$$= P(B) \times P(R)$$
$$= \frac{4}{7} \times \frac{3}{7} = \frac{12}{49}$$ The AND rule

P(red and blue)
$$= P(R) \times P(B)$$
$$= \frac{3}{7} \times \frac{4}{7} = \frac{12}{49}$$ The AND rule

P(one of either colour)
$$= \frac{12}{49} + \frac{12}{49} = \frac{24}{49}$$ The OR rule

the counters in the example had not been replaced,
would be an example of **conditional probability**.
his is when the outcome of the second event is
ependent on the outcome of the first. The tree
iagram for the example without replacement would
ook like this:

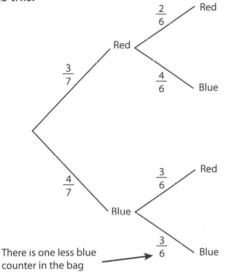

There is one less blue
counter in the bag

Probability of picking two blues:

$$P(B) \times P(B)$$

$$= \frac{4}{7} \times \frac{3}{6} = \frac{12}{42}$$

$$= \frac{2}{7}$$

SUMMARY

- If two events are mutually exclusive, the
  probability of A or B happening is found by
  adding the probabilities.

  $$P(A \text{ or } B) = P(A) + P(B)$$

- If two events are independent, the
  probability of A and B happening is found
  by multiplying the probabilities.

  $$P(A \text{ and } B) = P(A) \times P(B)$$

- These rules also work for more than two events.
- Remember that the probabilities on the
  branches leaving each point on a tree
  diagram should add up to 1.

QUESTIONS

QUICK TEST

1. Sangeeta has a biased dice. The probability of
   getting a three is 0.4. She rolls the dice twice.

   a. Complete the tree diagram.

   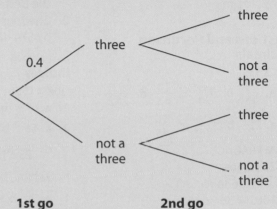

   1st go   **2nd go**

   b. Work out the probability that she gets:

      i. two threes

      ii. exactly one three

EXAM PRACTICE

1. A bag contains 3 red, 4 blue and 2 green beads.
   A bead is picked out of the bag at random and
   its colour noted. It is **not** replaced in the bag.
   A second bead is picked out at random. Work
   out the probability that two different-coloured
   beads are chosen.

2. On her way to work Mrs Harris drives down a
   country lane. The probability that she meets a
   tractor in the lane is 0.2. If she meets a tractor
   the probability that she is late for work is 0.6. If
   she does not meet a tractor the probability that
   she is late for work is 0.1.

   a. What is the probability that Mrs Harris
      meets a tractor and is late for work?

   b. What is the probability that Mrs Harris is not
      late for work?

Answers

Day 1

pages 4–5
Prime Factors, HCF and LCM
QUICK TEST
1. a. $50 = 2 \times 5^2$ b. $360 = 2^3 \times 3^2 \times 5$
   c. $16 = 2^4$
2. a. False b. True c. True d. False

EXAM PRACTICE
1. HCF $= 6$
2. 12.20 pm

pages 6–7
Fractions and Recurring Decimals
QUICK TEST
1. a. $\frac{13}{15}$ b. $2\frac{11}{21}$ c. $\frac{10}{63}$ d. $\frac{81}{242}$
2. a. $\frac{7}{9}$ b. $\frac{215}{999}$ c. $\frac{16}{45}$

EXAM PRACTICE
1. 28 pages
2. $x = 0.363636\ldots$ ①
   $100x = 36.363636\ldots$ ②
   ② − ① $99x = 36$
   $x = \frac{36}{99} = \frac{4}{11}$

pages 8–9
Repeated Percentage Change
QUICK TEST
1. 51.7% (3sf)   2. £121 856

EXAM PRACTICE
1. Savvy Saver – interest earned is £150, Money Grows – interest earned is £151.88
   Shamil is not correct – he would earn £1.88 more with the Money Grows investment.

pages 10–11
Reverse Percentage Problems
QUICK TEST
1. a. £57.50 b. £127
   c. £237.50 d. £437.50

EXAM PRACTICE
1. £250
2. Yes, Joseph is correct since $\frac{60}{0.85}$
   $= £70.59$

pages 12–13
Ratio and Proportion
QUICK TEST
1. £20 : £40 : £100 2. £35.28 3. 3 days

EXAM PRACTICE
1. There are many ways to work this out – this shows one possible way. Work out the cost of 25 ml for each tube of toothpaste.
   50 ml = £1.24; 25 ml = 62p
   75 ml = £1.96; 25 ml = 65.3̇p
   100 ml = £2.42; 25 ml = 60.5p
   The 100 ml tube of toothpaste is the better value for money.
2. Cheaper in America by £9.72 (or by $14.48).

pages 14–15
Indices
QUICK TEST
1. a. 6^8 b. 12^{13} c. 5^6 d. 4^2
2. a. $6b^{10}$ b. $2b^{-16}$ c. $9b^8$
   d. $\frac{1}{25x^4y^6}$ or $\frac{1}{25}x^{-4}y^{-6}$

EXAM PRACTICE
1. a. 1   b. $\frac{1}{49}$ c. 4 d. $\frac{1}{9}$
2. a. $x^{-4} = \frac{1}{x^4}$ b. $6x^3$

Day 2

pages 16–17
Standard Index Form
QUICK TEST
1. a. 6.4×10^4 b. 4.6×10^{-4}
2. a. 1.2×10^{11} b. 2×10^{-1}
3. a. 1.4375×10^{18} b. 5.476×10^{19}

EXAM PRACTICE
1. a. 4×10^7 b. 0.00006
2. 1.4×10^{-6} g

pages 18–19
Proportionality
QUICK TEST
1. a. $y = kx$ b. $y = \frac{k}{\sqrt[3]{x}}$
   c. $y = \frac{k}{x}$ d. $y = kx^3$

EXAM PRACTICE
1. $a = k\sqrt{x}$
   $8 = k\sqrt{4}$
   $\therefore k = 4$
   $a = 4\sqrt{x}$
   $64 = 4\sqrt{x}$ $\therefore x = 256$
2. a. $y = \frac{90}{4}$ $y = 22.5$ b. $x = \pm3.87$

pages 20–21
Surds
QUICK TEST
1. a. $2\sqrt{6}$ b. $10\sqrt{2}$ c. $6\sqrt{3}$
2. Molly is incorrect since
   $(2 - \sqrt{3})^2 = (2 - \sqrt{3})(2 - \sqrt{3})$
   $= 4 - 4\sqrt{3} + 3$
   $= 7 - 4\sqrt{3}$
3. Correct since $\frac{1}{\sqrt{2}}$ has been rationalised to give $\frac{\sqrt{2}}{2}$

EXAM PRACTICE
1. $\frac{5 - \sqrt{75}}{\sqrt{3}} \times \frac{\sqrt{3}}{\sqrt{3}}$
   $= \frac{5\sqrt{3} - \sqrt{225}}{3} = \frac{5\sqrt{3}}{3} - 5$
   $= -5 + \frac{5\sqrt{3}}{3}$
   $\therefore a = -5, \ b = \frac{5}{3}$
2. $\frac{3}{\sqrt{6}} \times \frac{\sqrt{6}}{\sqrt{6}} = \frac{3\sqrt{6}}{6} = \frac{\sqrt{6}}{2}$
3. $\frac{(5 + \sqrt{5})(2 - 2\sqrt{5})}{\sqrt{45}}$
   $= \frac{10 - 10\sqrt{5} + 2\sqrt{5} - 2(\sqrt{5})^2}{\sqrt{45}}$
   $= \frac{10 - 8\sqrt{5} - 10}{\sqrt{45}} = \frac{-8\sqrt{5}}{3\sqrt{5}} = -\frac{8}{3}$

pages 22–23
Upper and Lower Bounds
QUICK TEST
1. A – upper bound E – lower bound
2. Upper bound $= 0.2\dot{2}\dot{6}$
   Lower bound $= 0.2\dot{1}\dot{8}$

EXAM PRACTICE
1. 0.0437 (3sf)

pages 24–25
Formulae and Expressions
QUICK TEST
1. a. $8a - 7b$ b. $4a^2 - 8b^2$
   c. $2xy + 2xy^2$
2. a. $-\frac{31}{5} = -6\frac{1}{5}$ b. 4.36 or $4\frac{9}{25}$ c. 9
3. $u = \pm\sqrt{v^2 - 2as}$
4. $p = \frac{-(t + vq)}{q - 1} = \frac{(t + vq)}{1 - q}$

EXAM PRACTICE
1. a. Josh is correct since $3x^2$ means $3 \times x^2$, this gives $3 \times 2^2 = 12$
   b. 36

c. $x = \pm\sqrt{\dfrac{y}{4}} + 1$ **or** $x = \pm\dfrac{\sqrt{y}}{2} + 1$

pages 26–27

Brackets and Factorisation

QUICK TEST

1. a. $x^2 + x - 6$ **b.** $4x^2 - 12x$
c. $x^2 - 6x + 9$
2. a. $6x(2y - x)$ **b.** $3ab\,(a + 2b)$
c. $(x + 2)(x + 2) = (x + 2)^2$
d. $(x + 1)(x - 5)$
e. $(x - 10)(x + 10)$

EXAM PRACTICE

1. a. $3t^2 - 4t$ **b.** $6x + 4$
2. a. $y(y + 1)$ **b.** $5pq(p - 2q)$
c. $(a + b)(a + b + 4)$
d. $(x - 2)(x - 3)$

Day 3

pages 28–29

Equations 1

QUICK TEST

1. $x = 8$ **2.** $x = -5$ **3.** $x = -\dfrac{1}{2}$
4. $x = -3.25$ **5.** $x = -\dfrac{1}{2}$ **6.** $x = 17$

EXAM PRACTICE

1. a. $x = 2.4$ **b.** $x = -2.5$ **c.** $y = 2$
2. a. $x = -\dfrac{1}{5}$ **b.** $x = -4\dfrac{1}{3}$

pages 30–31

Equations 2

QUICK TEST

1. $x = 5.5$ cm, shortest length is
$2x - 5 = 6$ cm
2. $k + 3 = \dfrac{1}{3}$ $k = -\dfrac{8}{3}$
3. $4^{2k} = 4^3$ $2k = 3$ $k = \dfrac{3}{2}$
4. $2^{3k-1} = 2^6$ $3k - 1 = 6$ $k = \dfrac{7}{3}$

EXAM PRACTICE

1. $x = 54$, smallest angle $= 64°$
2. $16^{2k} = 64^{k+1}$
$(2^4)^{2k} = (2^6)^{k+1}$
$2^{8k} = 2^{6k+6}$
$8k = 6k + 6$
$2k = 6$
$k = 3$

pages 32–33

Solving Quadratic Equations

QUICK TEST

1. a. $(x - 2)(x - 4) = 0$ $x = 2, x = 4$
b. $(x + 4)(x + 1) = 0$ $x = -4, x = -1$
c. $(x + 2)(x - 6) = 0$ $x = -2, x = 6$

2. a. $x = -5, x = 3$ **b.** $x = -\dfrac{1}{2}, x = -2$

EXAM PRACTICE

1. a. Shaded area
$= (2x + 5)(x + 3) - (x + 1)^2$
$= (2x^2 + 11x + 15) - (x^2 + 2x + 1)$
Area $= x^2 + 9x + 14$
Since the shaded area $= 45\,\text{cm}^2$
$45 = x^2 + 9x + 14$
So $x^2 + 9x + 14 - 45 = 0$
$x^2 + 9x - 31 = 0$
b.i. $x = \dfrac{-b \pm \sqrt{b^2 - 4ac}}{2a}$

$x = \dfrac{-9 \pm \sqrt{9^2 - (4 \times 1 \times -31)}}{2 \times 1}$

$x = 2.66\,(3\text{sf})$ or $x = -11.7\,(3\text{sf})$
ii. Reject $x = -11.7$
Perimeter $3.66... \times 4 =$
14.6 cm (3sf)

pages 34–35

Solving Quadratic Equations and Cubic Equations

QUICK TEST

1. 3.3 (1dp)
2. $p = -2, q = 3$

EXAM PRACTICE

1. $x = 2.7$

x	$x^3 + 4x^2 = 49$	Comment
2	$2^3 + 4 \times 2^2 = 24$	too small
3	$3^3 + 4 \times 3^2 = 63$	too big
2.5	$2.5^3 + 4 \times 2.5^2 = 40.625$	too small
2.7	$2.7^3 + 4 \times 2.7^2 = 48.843$	too small
2.75	$2.75^3 + 4 \times 2.75^2 = 51.046...$	too big
2.74	$2.74^3 + 4 \times 2.74^2 = 50.601...$	too big

2. a. $(x + 5)^2 - 20$ $\therefore a = 5, b = -20$
b. -20

pages 36–37

Simultaneous Linear Equations

QUICK TEST

1. $a = 4$ $b = -4.5$ **2.** $x = 2, y = 4$

EXAM PRACTICE

1. $a = 3$ $b = -2$ **2.** Hat £2, Balloon £3

pages 38–39

Solving a Linear and a Non-linear Equation Simultaneously

QUICK TEST

1. a. $x = -1, y = 3$ $x = 2, y = 6$

b. $x = 3, y = 4$ $x = -4, y = -3$
2. a. This is where the quadratic
graph $y = x^2 + 2$ intersects the
straight line graph $y = x + 4$.
Their points of intersection are
$(-1, 3)$ and $(2, 6)$.
b. This is where the circle
$x^2 + y^2 = 25$ intersects the
straight line $y = x + 1$. Their
points of intersection are $(3, 4)$
and $(-4, -3)$.

EXAM PRACTICE

1. $(0, 5)$, $(-5, 0)$
2. $x = 1, y = 5$ or $x = -3.4, y = -3.8$

Day 4

pages 40–41

Algebraic Fractions

QUICK TEST

1. a. no **b.** yes **c.** yes
2. a. $\dfrac{5x - 1}{(x + 1)(x - 1)}$ **b.** $\dfrac{2aq}{3c}$

c. $\dfrac{40(a + b)}{3}$

EXAM PRACTICE

1. $\dfrac{8x + 16}{x^2 - 4} = \dfrac{8(x + 2)}{(x + 2)(x - 2)} = \dfrac{8}{x - 2}$

2. $\dfrac{x^2(6 + x)}{x^2 - 36} = \dfrac{x^2(6 + x)}{(x + 6)(x - 6)} = \dfrac{x^2}{x - 6}$

3. $\dfrac{2x^2 + 7x - 15}{x^2 + 3x - 10} = \dfrac{(2x - 3)(x + 5)}{(x + 5)(x - 2)}$
$= \dfrac{2x - 3}{x - 2}$

4. $\dfrac{3}{x} - \dfrac{4}{x + 1} = 1$

$\dfrac{3(x + 1) - 4x}{x(x + 1)} = 1$

$3x + 3 - 4x = x^2 + x$

$x^2 + 2x - 3 = 0$

$(x - 1)(x + 3) = 0$

$x = 1$ or $x = -3$

pages 42–43

Inequalities

QUICK TEST

1. a. $x < 2.2$ **b.** $\dfrac{4}{3} \leqslant x < 3$

c. $x > -\dfrac{9}{5}$

1. a. −2, −1, 0, 1, 2, 3, 4   **b.** $x < 2$
2.

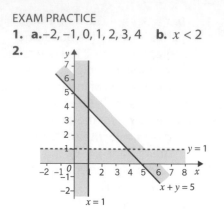

pages 44–45
Straight-line Graphs
QUICK TEST
1. a.

x	−2	−1	0	1	2	3
y	−1	1	3	5	7	9

b.

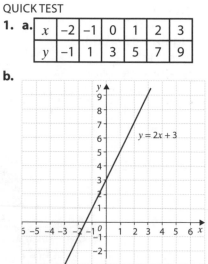

EXAM PRACTICE
1. a. $y = -2x + 3$ **b.** $(\frac{1}{2}, 1)$
   c. $y = \frac{1}{2}x + 3$

pages 46–47
Curved Graphs
QUICK TEST
1. Graph A: $y = \frac{3}{x}$

   Graph B: $y = 4x + 2$
   Graph C: $y = x^3 - 5$
   Graph D: $y = 2 - x^2$
EXAM PRACTICE
1. a.

x	−3	−2	−1	0	1	2	3
y	−28	−9	−2	−1	0	7	26

b.

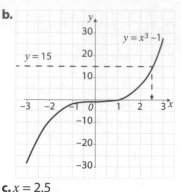

c. $x = 2.5$

pages 48–49
Interpreting Graphs
QUICK TEST
1. True for both solutions
2. False for both solutions
3. True for $x = -0.7$, false for $x = 5.6$
EXAM PRACTICE
1. $a = 9000$ $b = \frac{2}{3}$

pages 50–51
Functions and Transformations
QUICK TEST
1. a.

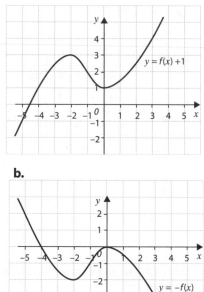

b.

EXAM PRACTICE
1. a. (−1, −3) **b.** (2, −7) **c.** (4, −3)
   d. (−2, −3) **e.** (1, −3)

Day 5
pages 52–53
Loci

QUICK TEST
1.

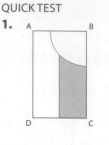

EXAM PRACTICE
1. a.b.

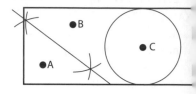

pages 54–55
Translations and Reflections
QUICK TEST
1. a. Reflection in $y = 0$ (x-axis)
   b. Reflection in $x = 0$ (y-axis)
   c. Translation of $\begin{pmatrix} -6 \\ 0 \end{pmatrix}$
   d. Translation of $\begin{pmatrix} 5 \\ -1 \end{pmatrix}$
EXAM PRACTICE
1. a. b. See below

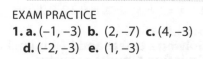

c. Reflection in the line $y = x$

pages 56–57
Rotation and Enlargement
QUICK TEST
1.

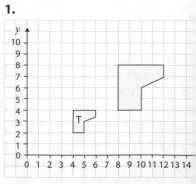

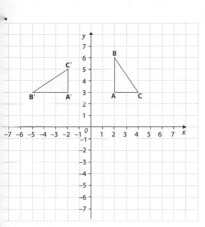

XAM PRACTICE

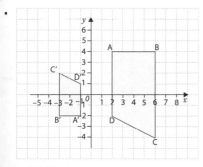

ages 58–59
Similarity and Congruency 1
QUICK TEST
. **a.** 20.9 cm **b.** 13.8 cm **c.** 4.5 cm
. Yes they are congruent since
76° + 48° + 56° = 180°. Hence
both triangles have two sides
the same and the included
angle of 56° (SAS)
XAM PRACTICE
. 20.7 cm (3sf)
. There are several ways of proving
this. An example might be:
Since triangle CDE is equilateral,
length CD = length CE. Both
triangles have a common
length, CF, which is the
perpendicular bisector of DE.

Hence length DF = length EF.

By SSS triangle CFD is congruent
to triangle CFE.

ages 60 61
Similarity and Congruency 2
QUICK TEST
. 28.4 cm² (3sf)   **2.** 78.75 cm²
. 2.1875 cm³

EXAM PRACTICE
1. 1890 g   **2.** 58 320 cm³

pages 62–63
Circle Theorems
QUICK TEST
1. a. 62° **b.** 109° **c.** 53° **d.** 50° **e.** 126°
EXAM PRACTICE
1. a. 19° A tangent and radius
meet at 90°
b. 71° Alternate Segment
Theorem or angle in a
semicircle is 90°
$\hat{EFG} = 90°$. Hence $\hat{GEF} = 71°$.

pages 64–65
Pythagoras' Theorem
QUICK TEST
1. a. 17.46 cm (2dp) **b.** 9.38 cm (2dp)
EXAM PRACTICE
1. Since $26^2 = 24^2 + 10^2$
676 = 576 + 100. Since this obeys
Pythagoras' Theorem, then the
triangle must be right-angled.
2. £15.66
3. $\sqrt{149} = 12.2$

Day 6
pages 66–67
Trigonometry
QUICK TEST
1. a. 5.79 cm   **b.** 8.40 cm
2. a. 38.7°   **b.** 43.0°
EXAM PRACTICE
1. a. 57.4° (3sf)
b. $\sin 48° = \dfrac{19.7}{PA}$ $PA = \dfrac{19.7}{\sin 48°}$
$PA = 26.5$ m (3sf)

pages 68–69
Trigonometry in 3D
QUICK TEST
1. a. i. True   **ii.** False   **iii.** False
iv. True
b. 10° (nearest degree)
EXAM PRACTICE
1. a. 19.18 cm (2dp)   **b.** 73.6° (3sf)

pages 70–71
Sine and Cosine Rules
QUICK TEST
1. a. Correct   **b.** Correct
2. a. 114° (nearest degree)

---

b. 53° (nearest degree)
EXAM PRACTICE
1. 16.4 m (3sf)

pages 72–73
Trigonometric Functions
QUICK TEST
1.

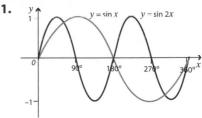

2. 60°, 300°
EXAM PRACTICE
1. −315°, −225°, 45°, 135°
2. 30°, 330°

pages 74–75
Arc, Sector and Segment
QUICK TEST
1. a. 14.05 cm
b. 49.17 cm²
c. 26.97 cm²
2. True since
$\left(\dfrac{70°}{360°} \times \pi \times 8^2 - \dfrac{1}{2} \times 8 \times 8 \times \sin 70°\right)$
= 9.03 cm²
EXAM PRACTICE
1. 30 cm²

pages 76–77
Surface Area and Volume
QUICK TEST
1. a. 565 cm³
b. 637 cm³
c. 905 cm³
d. 245 cm³
EXAM PRACTICE
1. Radius = 2.61 m (3sf)
2. $\pi \times (2x)^2 \times h = \dfrac{4}{3} \times \pi \times (3x)^3$

$4x^2 \times \pi \times h = \dfrac{4}{3} \times \pi \times 27x^3$

$4\pi x^2 h = 36\pi x^3$

$h = \dfrac{36\pi x^3}{4\pi x^2}$

$h = 9x$

pages 78–79
Vectors 1
QUICK TEST

1.

b

2. Yes, vectors **c** and **d** are parallel since **d** = 2**c**

3. a. $\sqrt{4^2+2^2} = \sqrt{20} = 2\sqrt{5}$
   b. $\sqrt{(-3)^2+4^2}\ \sqrt{25} = 5$
   c. $\sqrt{(-2)^2+0^2} = \sqrt{4} = 2$
   d. $\sqrt{6^2+(-1)^2} = \sqrt{37}$

EXAM PRACTICE

1. a. $\begin{pmatrix} 4 \\ 5 \end{pmatrix}$
   b. (5, 4)

Day 7

pages 80–81
Vectors 2
QUICK TEST

1. a. i. $-\mathbf{a}+\mathbf{c}$ ii. $-\mathbf{c}-\mathbf{a}$
   b. $\mathbf{a}+\frac{1}{2}(-\mathbf{a}+\mathbf{c}) = \frac{1}{2}\mathbf{a}+\frac{1}{2}\mathbf{c}$
   or $\frac{1}{2}(\mathbf{a}+\mathbf{c})$

EXAM PRACTICE

1. a. i. $-\mathbf{a}+\mathbf{b}$ ii. $-2\mathbf{a}+2\mathbf{b}$
   b. $-\mathbf{a}+\mathbf{b}$
   c. Since $\overrightarrow{FE} = -\mathbf{a}+\mathbf{b}$
      and $\overrightarrow{AD} = 2(-\mathbf{a}+\mathbf{b}) = 2\overrightarrow{FE}$
      So AD is twice the length of and parallel to FE.

pages 82–83
Scatter Diagrams and Correlation
QUICK TEST

1. True   2. False   3. True
4. True   5. True

EXAM PRACTICE

1. a. Line of best fit should be drawn as close as possible to all points and in the direction of the data.
   b. approx. 2 years old
   c. approx. £3000

pages 84–85
Averages
QUICK TEST

1. 150.9

2. a.

```
0 | 7  9
1 | 8
2 | 5  5  5  6  7  7  8
3 | 1  1  1  3  6  6  7  9
4 | 0  2  2  7  9
5 | 0
```
   Key 4|2 = 42

   b. Median = 31

EXAM PRACTICE

1. 5.52 hours

pages 86–87
Cumulative Frequency Graphs
QUICK TEST

1.

	Time in Minutes
Median time	17
Lower quartile	12
Interquartile range	11
Longest time	29

EXAM PRACTICE

1. a. 9B
   b. Class 9A. The median is higher than 9B: 50% of students in 9A scored over 65 marks. The top 50% of students in class 9B scored over 55 marks.

pages 88–89
Histograms
QUICK TEST

1.
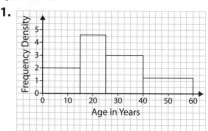

EXAM PRACTICE

1. a.

Distance (km)	Frequency	Frequency Density
$0 < d \leqslant 15$	12	0.8
$15 < d \leqslant 25$	32	3.2
$25 < d \leqslant 30$	36	7.2
$30 < d \leqslant 45$	60	4.0
$45 < d \leqslant 55$	20	2.0

b.
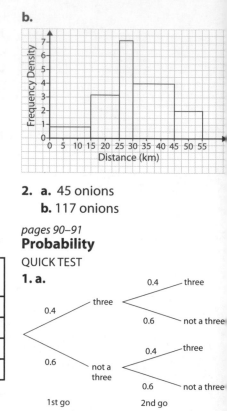

2. a. 45 onions
   b. 117 onions

pages 90–91
Probability
QUICK TEST

1. a.
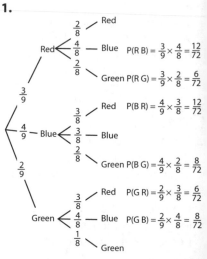

   b. i. 0.16   ii. 0.48

EXAM PRACTICE

1.

P(R B) = $\frac{3}{9}\times\frac{4}{8} = \frac{12}{72}$
Green P(R G) = $\frac{3}{9}\times\frac{2}{8} = \frac{6}{72}$
P(B R) = $\frac{4}{9}\times\frac{3}{8} = \frac{12}{72}$
Green P(B G) = $\frac{4}{9}\times\frac{2}{8} = \frac{8}{72}$
P(G R) = $\frac{2}{9}\times\frac{3}{8} = \frac{6}{72}$
P(G B) = $\frac{2}{9}\times\frac{4}{8} = \frac{8}{72}$

P(two different colours) = $\frac{52}{72} = \frac{13}{18}$
Or 1 – P(same colours)
$= 1 - \left(\left(\frac{3}{9}\times\frac{2}{8}\right)+\left(\frac{4}{9}\times\frac{3}{8}\right)+\left(\frac{2}{9}\times\frac{1}{8}\right)\right)$
$= 1 - \frac{20}{72} = \frac{52}{72} = \frac{13}{18}$

2. a. 0.12   b. 0.8